STUDENT SOLUTIONS MANUAL

SHARON MYERS
Radford University

KEYING YE
University of Texas at San Antonio

EIGHTH EDITION

PROBABILITY & STATISTICS

for Engineers & Scientists

WALPOLE   MYERS   MYERS   YE

PEARSON

Prentice Hall

Upper Saddle River, NJ 07458

Editor-in-Chief: Sally Yagan
Executive Acquisitions Editor: Petra Recter
Supplement Editor: Joanne Wendelken
Executive Managing Editor: Kathleen Schiaparelli
Senior Managing Editor: Nicole M. Jackson
Assistant Managing Editor: Karen Bosch Petrov
Production Editor: Traci Douglas
Supplement Cover Manager: Paul Gourhan
Supplement Cover Designer: Christopher Kossa
Manufacturing Buyer: Ilene Kahn
Manufacturing Manager: Alexis Heydt-Long

© 2007 Pearson Education, Inc.
Pearson Prentice Hall
Pearson Education, Inc.
Upper Saddle River, NJ 07458

Printed in the United States of America

10 9 8 7 6 5 4 3 2

ISBN 0-13-187713-5

Pearson Education Ltd., *London*
Pearson Education Australia Pty. Ltd., *Sydney*
Pearson Education Singapore, Pte. Ltd.
Pearson Education North Asia Ltd., *Hong Kong*
Pearson Education Canada, Inc., *Toronto*
Pearson Educación de Mexico, S.A. de C.V.
Pearson Education—Japan, *Tokyo*
Pearson Education Malaysia, Pte. Ltd.

Contents

Chapter 1

Introduction to Statistics and Data Analysis

1.1 (a) 15.

(b) $\bar{x} = \frac{1}{15}(3.4 + 2.5 + 4.8 + \cdots + 4.8) = 3.787$.

(c) Sample median is the 8th value, after the data is sorted from smallest to largest: 3.6.

(d) A dot plot is shown below.

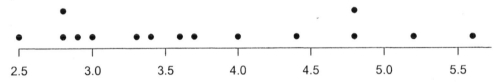

(e) After trimming total 40% of the data (20% highest and 20% lowest), the data becomes:

$$2.9 \quad 3.0 \quad 3.3 \quad 3.4 \quad 3.6$$
$$3.7 \quad 4.0 \quad 4.4 \quad 4.8$$

So. the trimmed mean is

$$x_{\text{tr}20} = \frac{1}{9}(2.9 + 3.0 + \cdots + 4.8) = 3.678.$$

1.3 (a) A dot plot is shown below.

```
        o    o o o       o   o     o o × × o × o    × ×      ×    × × ×
    ┼──────────┼──────────┼──────────┼──────────┼──────────┼──────────┼
   200        205        210        215        220        225        230
```

In the figure, "×" represents the "No aging" group and "o" represents the "Aging" group.

(b) Yes; tensile strength is greatly reduced due to the aging process.

(c) $\text{Mean}_{\text{Aging}} = 209.90$, and $\text{Mean}_{\text{No aging}} = 222.10$.

1

(d) Median$_{\text{Aging}}$ = 210.00, and Median$_{\text{No aging}}$ = 221.50. The means and medians for each group are similar to each other.

1.5 (a) A dot plot is shown below.

In the figure, "×" represents the control group and "o" represents the treatment group.

(b) $\bar{X}_{\text{Control}}$ = 5.60, $\tilde{X}_{\text{Control}}$ = 5.00, and $\bar{X}_{\text{tr(10);Control}}$ = 5.13;
$\bar{X}_{\text{Treatment}}$ = 7.60, $\tilde{X}_{\text{Treatment}}$ = 4.50, and $\bar{X}_{\text{tr(10);Treatment}}$ = 5.63.

(c) The difference of the means is 2.0 and the differences of the medians and the trimmed means are 0.5, which are much smaller. The possible cause of this might be due to the extreme values (outliers) in the samples, especially the value of 37.

1.7 $s^2 = \frac{1}{15-1}[(3.4-3.787)^2+(2.5-3.787)^2+(4.8-3.787)^2+\cdots+(4.8-3.787)^2] = 0.94284$;
$s = \sqrt{s^2} = \sqrt{0.9428} = 0.971$.

1.9 $s^2_{\text{No Aging}} = \frac{1}{10-1}[(227-222.10)^2+(222-222.10)^2+\cdots+(221-222.10)^2] = 42.12$;
$s_{\text{No Aging}} = \sqrt{42.12} = 6.49$.
$s^2_{\text{Aging}} = \frac{1}{10-1}[(219-209.90)^2+(214-209.90)^2+\cdots+(205-209.90)^2] = 23.62$;
$s_{\text{Aging}} = \sqrt{23.62} = 4.86$.

1.11 For the control group: s^2_{Control} = 69.39 and s_{Control} = 8.33.
For the treatment group: $s^2_{\text{Treatment}}$ = 128.14 and $s_{\text{Treatment}}$ = 11.32.

1.13 (a) Mean = $\bar{X}$ = 124.3 and median = $\tilde{X}$ = 120;

(b) 175 is an extreme observation.

1.15 Yes. The value 0.03125 is actually a *P*-value and a small value of this quantity means that the outcome (i.e., $HHHHH$) is very unlikely to happen with a fair coin.

1.17 (a) $\bar{X}_{\text{smokers}}$ = 43.70 and $\bar{X}_{\text{nonsmokers}}$ = 30.32;

(b) s_{smokers} = 16.93 and $s_{\text{nonsmokers}}$ = 7.13;

(c) A dot plot is shown below.

In the figure, "×" represents the nonsmoker group and "o" represents the smoker group.

(d) Smokers appear to take longer time to fall asleep and the time to fall asleep for smoker group is more variable.

1.19 (a) A stem-and-leaf plot is shown below.

Stem	Leaf	Frequency
0	22233457	8
1	023558	6
2	035	3
3	03	2
4	057	3
5	0569	4
6	0005	4

(b) The following is the relative frequency distribution table.

Relative Frequency Distribution of Years

Class Interval	Class Midpoint	Frequency, f	Relative Frequency
$0.0 - 0.9$	0.45	8	0.267
$1.0 - 1.9$	1.45	6	0.200
$2.0 - 2.9$	2.45	3	0.100
$3.0 - 3.9$	3.45	2	0.067
$4.0 - 4.9$	4.45	3	0.100
$5.0 - 5.9$	5.45	4	0.133
$6.0 - 6.9$	6.45	4	0.133

(c) $\bar{X} = 2.797$, $s = 2.227$ and Sample range is $6.5 - 0.2 = 6.3$.

1.21 (a) $\bar{X} = 1.7743$ and $\tilde{X} = 1.7700$;

(b) $s = 0.3905$.

1.23 (a) A dot plot is shown next.

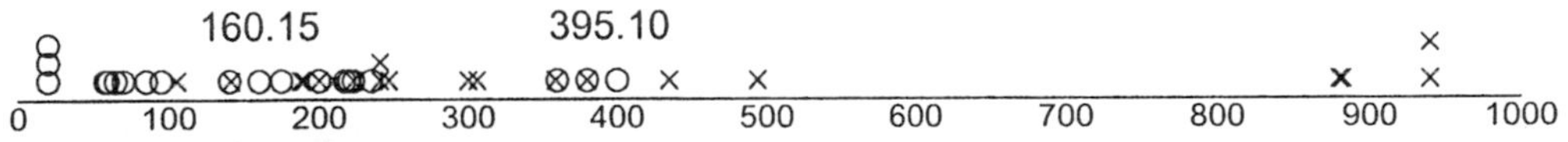

(b) $\bar{X}_{1980} = 395.1$ and $\bar{X}_{1990} = 160.2$.

(c) The sample mean for 1980 is over twice as large as that of 1990. The variability for 1990 decreased also as seen by looking at the picture in (a). The gap represents an increase of over 400 ppm. It appears from the data that hydrocarbon emissions decreased considerably between 1980 and 1990 and that the extreme large emission (over 500 ppm) were no longer in evidence.

1.25 (a) $\bar{X} - 33.31$,

(b) $\tilde{X} = 26.35$;

(c) A histogram plot is shown next.

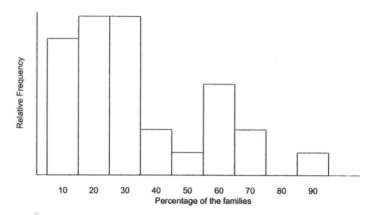

(d) $\bar{X}_{tr(10)} = 30.97$. This trimmed mean is in the middle of the mean and median using the full amount of data. Due to the skewness of the data to the right (see plot in (c)), it is common to use trimmed data to have a more robust result.

1.27 (a) The averages of the wear are plotted here.

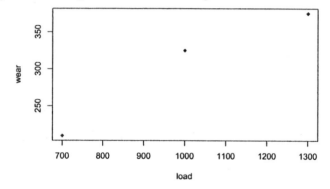

(b) When the load value increases, the wear value also increases. It does show certain relationship.

(c) A plot of wears is shown next.

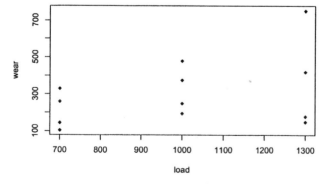

(d) The relationship between load and wear in (c) is not as strong as the case in (a), especially for the load at 1300. One reason is that there is an extreme value (750) which influence the mean value at the load 1300.

1.29 (a) A dot plot is shown next.

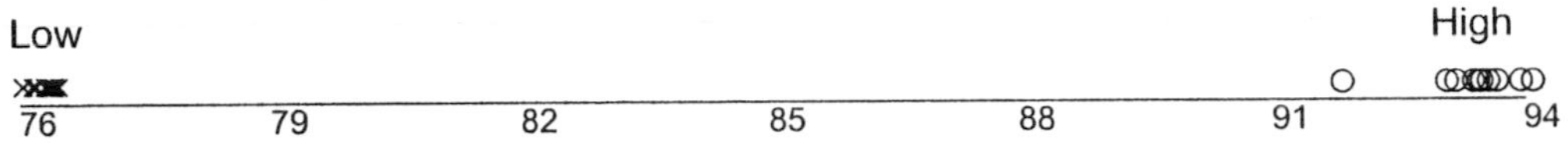

Low ... High

In the figure, "×" represents the low-injection-velocity group and "∘" represents the high-injection-velocity group.

(b) In this time, the shrinkage values are much higher for the high-injection-velocity group than those for the low-injection-velocity group. Also, the variation for the former group is much higher as well.

(c) Since the shrinkage effects change in different direction between low mode temperature and high mold temperature, the apparent interactions between the mold temperature and injection velocity are significant.

Chapter 2

Probability

2.1 (a) $S = \{8, 16, 24, 32, 40, 48\}$.

    (b) For $x^2 + 4x - 5 = (x + 5)(x - 1) = 0$, the only solutions are $x = -5$ and $x = 1$.
        $S = \{-5, 1\}$.

    (c) $S = \{T, HT, HHT, HHH\}$.

    (d) $S = \{\text{N. America, S. America, Europe, Asia, Africa, Australia, Antarctica}\}$.

    (e) Solving $2x - 4 \geq 0$ gives $x \geq 2$. Since we must also have $x < 1$, it follows that
        $S = \phi$.

2.3 (a) $A = \{1, 3\}$.

    (b) $B = \{1, 2, 3, 4, 5, 6\}$.

    (c) $C = \{x \mid x^2 - 4x + 3 = 0\} = \{x \mid (x - 1)(x - 3) = 0\} = \{1, 3\}$.

    (d) $D = \{0, 1, 2, 3, 4, 5, 6\}$. Clearly, $A = C$.

2.5 $S = \{1HH, 1HT, 1TH, 1TT, 2H, 2T, 3HH, 3HT, 3TH, 3TT, 4H, 4T, 5HH, 5HT, 5TH,$
    $5TT, 6H, 6T\}$.

2.7 $S_1 = \{MMMM, MMMF, MMFM, MFMM, FMMM, MMFF, MFMF, MFFM,$
    $FMFM, FFMM, FMMF, MFFF, FMFF, FFMF, FFFM, FFFF\}$.
    $S_2 = \{0, 1, 2, 3, 4\}$.

2.9 (a) $A = \{1HH, 1HT, 1TH, 1TT, 2H, 2T\}$.

    (b) $B = \{1TT, 3TT, 5TT\}$.

    (c) $A' = \{3HH, 3HT, 3TH, 3TT, 4H, 4T, 5HH, 5HT, 5TH, 5TT, 6H, 6T\}$.

    (d) $A' \cap B = \{3TT, 5TT\}$.

    (e) $A \cup B = \{1HH, 1HT, 1TH, 1TT, 2H, 2T, 3TT, 5TT\}$.

2.11 (a) $S = \{M_1 M_2, M_1 F_1, M_1 F_2, M_2 M_1, M_2 F_1, M_2 F_2, F_1 M_1, F_1 M_2, F_1 F_2, F_2 M_1, F_2 M_2,$
    $F_2 F_1\}$.

    (b) $A = \{M_1 M_2, M_1 F_1, M_1 F_2, M_2 M_1, M_2 F_1, M_2 F_2\}$.

(c) $B = \{M_1F_1, M_1F_2, M_2F_1, M_2F_2, F_1M_1, F_1M_2, F_2M_1, F_2M_2\}$.

(d) $C = \{F_1F_2, F_2F_1\}$.

(e) $A \cap B = \{M_1F_1, M_1F_2, M_2F_1, M_2F_2\}$.

(f) $A \cup C = \{M_1M_2, M_1F_1, M_1F_2, M_2M_1, M_2F_1, M_2F_2, F_1F_2, F_2F_1\}$.

(g)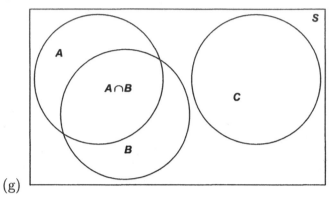

2.13 A Venn diagram is shown next.

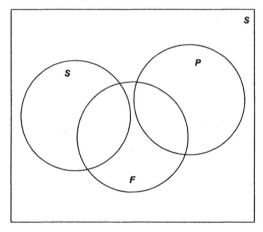

2.15 (a) $A' = \{\text{nitrogen, potassium, uranium, oxygen}\}$.

     (b) $A \cup C = \{\text{copper, sodium, zinc, oxygen}\}$.

     (c) $A \cap B' = \{\text{copper, zinc}\}$ and
        $C' = \{\text{copper, sodium, nitrogen, potassium, uranium zinc}\}$;
        so $(A \cap B') \cup C' = \{\text{copper, sodium, nitrogen, potassium, uranium, zinc}\}$.

     (d) $B' \cap C' = \{\text{copper, uranium, zinc}\}$.

     (e) $A \cap B \cap C = \phi$.

     (f) $A' \cup B' = \{\text{copper, nitrogen, potassium, uranium, oxygen, zinc}\}$ and
        $A' \cap C = \{\text{oxygen}\}$; so, $(A' \cup B') \cap (A' \cap C) = \{\text{oxygen}\}$.

2.17 A Venn diagram is shown next.

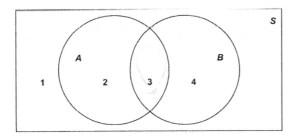

(a) From the above Venn diagram, $(A \cap B)'$ contains the regions of 1, 2 and 4.

(b) $(A \cup B)'$ contains region 1.

(c) A Venn diagram is shown next.

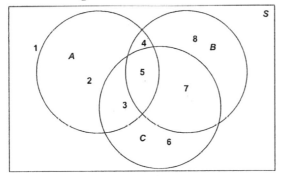

$(A \cap C) \cup B$ contains the regions of 3, 4, 5, 7 and 8.

2.19 (a) The family will experience mechanical problems but will receive no ticket for traffic violation and will not arrive at a campsite that has no vacancies.

(b) The family will receive a traffic ticket and arrive at a campsite that has no vacancies but will not experience mechanical problems.

(c) The family will experience mechanical problems and will arrive at a campsite that has no vacancies.

(d) The family will receive a traffic ticket but will not arrive at a campsite that has no vacancies.

(e) The family will not experience mechanical problems.

2.21 With $n_1 = 6$ sightseeing tours each available on $n_2 = 3$ different days, the multiplication rule gives $n_1 n_2 = (6)(3) = 18$ ways for a person to arrange a tour.

2.23 Since the die can land in $n_1 = 6$ ways and a letter can be selected in $n_2 = 26$ ways, the multiplication rule gives $n_1 n_2 = (6)(26) = 156$ points in S.

2.25 With $n_1 = 5$ different shoe styles in $n_2 = 4$ different colors, the multiplication rule gives $n_1 n_2 = (5)(4) = 20$ different pairs of shoes.

2.27 Using the generalized multiplication rule, there are $n_1 \times n_2 \times n_3 \times n_4 = (4)(3)(2)(2) = 48$ different house plans available.

2.29 With $n_1 = 3$ race cars, $n_2 = 5$ brands of gasoline, $n_3 = 7$ test sites, and $n_4 = 2$ drivers, the generalized multiplication rule yields $(3)(5)(7)(2) = 210$ test runs.

2.31 (a) With $n_1 = 4$ possible answers for the first question, $n_2 = 4$ possible answers for the second question, and so forth, the generalized multiplication rule yields $4^5 = 1024$ ways to answer the test.

   (b) With $n_1 = 3$ wrong answers for the first question, $n_2 = 3$ wrong answers for the second question, and so forth, the generalized multiplication rule yields

$$n_1 n_2 n_3 n_4 n_5 = (3)(3)(3)(3)(3) = 3^5 = 243$$

   ways to answer the test and get all questions wrong

2.33 Since the first digit is a 5, there are $n_1 = 9$ possibilities for the second digit and then $n_2 = 8$ possibilities for the third digit. Therefore, by the multiplication rule there are $n_1 n_2 = (9)(8) = 72$ registrations to be checked.

2.35 The first house can be placed on any of the $n_1 = 9$ lots, the second house on any of the remaining $n_2 = 8$ lots, and so forth. Therefore, there are $9! = 362,880$ ways to place the 9 homes on the 9 lots.

2.37 The first seat must be filled by any of 5 girls and the second seat by any of 4 boys. Continuing in this manner, the total number of ways to seat the 5 girls and 4 boys is $(5)(4)(4)(3)(3)(2)(2)(1)(1) = 2880$.

2.39 (a) Any of the $n_1 = 8$ finalists may come in first, and of the $n_2 = 7$ remaining finalists can then come in second, and so forth. By Theorem 2.3, there $8! = 40320$ possible orders in which 8 finalists may finish the spelling bee.

   (b) The possible orders for the first three positions are $_8P_3 = \frac{8!}{5!} = 336$.

2.41 By Theorem 2.4, $_6P_4 = \frac{6!}{2!} = 360$.

2.43 By Theorem 2.5, there are $4! = 24$ ways.

2.45 By Theorem 2.6, there are $\frac{8!}{3!2!} = 3360$.

2.47 By Theorem 2.7, there are $\binom{12}{7,3,2} = 7920$ ways.

2.49 By Theorem 2.8, there are $\binom{8}{3} = 56$ ways.

2.51 (a) Sum of the probabilities exceeds 1.

   (b) Sum of the probabilities is less than 1.

   (c) A negative probability.

   (d) Probability of both a heart and a black card is zero

2.53 $S = \{\$10, \$25, \$100\}$ with weights $275/500 = 11/20$, $150/500 = 3/10$, and $75/500 = 3/20$, respectively. The probability that the first envelope purchased contains less than $100 is equal to $11/20 + 3/10 = 17/20$.

2.55 Consider the events
S: industry will locate in Shanghai,
B: industry will locate in Beijing.

   (a) $P(S \cap B) = P(S) + P(B) - P(S \cup B) = 0.7 + 0.4 - 0.8 = 0.3$.

   (b) $P(S' \cap B') = 1 - P(S \cup B) = 1 - 0.8 = 0.2$.

2.57 (a) Since 5 of the 26 letters are vowels, we get a probability of $5/26$.

   (b) Since 9 of the 26 letters precede j, we get a probability of $9/26$.

   (c) Since 19 of the 26 letters follow g, we get a probability of $19/26$.

2.59 By Theorem 2.2, there are $N = (26)(25)(24)(9)(8)(7)(6) = 47,174,400$ possible ways to code the items of which $n = (5)(25)(24)(8)(7)(6)(4) = 4,032,000$ begin with a vowel and end with an even digit. Therefore, $\frac{n}{N} = \frac{10}{117}$.

2.61 Since there are 20 cards greater than 2 and less than 8, the probability of selecting two of these in succession is

$$\left(\frac{20}{52}\right)\left(\frac{19}{51}\right) = \frac{95}{663}.$$

2.63 (a) $\frac{\binom{4}{3}\binom{48}{2}}{\binom{52}{5}} = \frac{94}{54145}$.

   (b) $\frac{\binom{13}{4}\binom{13}{1}}{\binom{52}{5}} = \frac{143}{39984}$.

2.65 (a) $P(M \cup H) = 88/100 = 22/25$;

   (b) $P(M' \cap H') = 12/100 = 3/25$;

   (c) $P(H \cap M') = 34/100 = 17/50$.

2.67 (a) 0.32;

   (b) 0.68;

   (c) office or den.

2.69 $P(A) = 0.2$ and $P(B) = 0.35$

   (a) $P(A') = 1 - 0.2 = 0.8$;

   (b) $P(A' \cap B') = 1 - P(A \cup B) = 1 - 0.2 - 0.35 - 0.45$;

   (c) $P(A \cup B) = 0.2 + 0.35 = 0.55$.

2.71 (a) $0.12 + 0.19 = 0.31$;

     (b) $1 - 0.07 = 0.93$;

     (c) $0.12 + 0.19 = 0.31$.

2.73 (a) $P(C) = 1 - P(A) - P(B) = 1 - 0.990 - 0.001 = 0.009$;

     (b) $P(B') = 1 - P(B) = 1 - 0.001 = 0.999$;

     (c) $P(B) + P(C) = 0.01$.

2.75 (a) $1 - 0.95 - 0.002 = 0.048$;

     (b) $(\$25.00 - \$20.00) \times 10,000 = \$50,000$;

     (c) $(0.05)(10,000) \times \$5.00 + (0.05)(10,000) \times \$20 = \$12,500$.

2.77 (a) The probability that a convict who pushed dope, also committed armed robbery.

     (b) The probability that a convict who committed armed robbery, did not push dope.

     (c) The probability that a convict who did not push dope also did not commit armed robbery.

2.79 Consider the events:
    M: a person is a male;
    S: a person has a secondary education;
    C: a person has a college degree.

     (a) $P(M \mid S) = 28/78 = 14/39$;

     (b) $P(C' \mid M') = 95/112$.

2.81 (a) $P(M \cap P \cap H) = \frac{10}{68} = \frac{5}{34}$;

     (b) $P(H \cap M \mid P') = \frac{P(H \cap M \cap P')}{P(P')} = \frac{22-10}{100-68} = \frac{12}{32} = \frac{3}{8}$.

2.83 (a) 0.018;

     (b) $0.22 + 0.002 + 0.160 + 0.102 + 0.046 + 0.084 = 0.614$;

     (c) $0.102/0.614 = 0.166$;

     (d) $\frac{0.102+0.046}{0.175+0.134} = 0.479$.

2.85 Consider the events:
    H: husband watches a certain show,
    W: wife watches the same show.

     (a) $P(W \cap H) = P(W)P(H \mid W) = (0.5)(0.7) = 0.35$.

     (b) $P(W \mid H) = \frac{P(W \cap H)}{P(H)} = \frac{0.35}{0.4} = 0.875$.

     (c) $P(W \cup H) = P(W) + P(H) - P(W \cap H) = 0.5 + 0.4 - 0.35 = 0.55$.

2.87 Consider the events:

A: the vehicle is a camper,

B: the vehicle has Canadian license plates.

(a) $P(B \mid A) = \frac{P(A \cap B)}{P(A)} = \frac{0.09}{0.28} = \frac{9}{28}$.

(b) $P(A \mid B) = \frac{P(A \cap B)}{P(B)} = \frac{0.09}{0.12} = \frac{3}{4}$.

(c) $P(B' \cup A') = 1 - P(A \cap B) = 1 - 0.09 = 0.91$.

2.89 Consider the events:

A: the doctor makes a correct diagnosis,

B: the patient sues.

$P(A' \cap B) = P(A')P(B \mid A') = (0.3)(0.9) = 0.27$.

2.91 Consider the events:

A: the house is open,

B: the correct key is selected.

$P(A) = 0.4$, $P(A') = 0.6$, and $P(B) = \frac{\binom{1}{1}\binom{7}{2}}{\binom{8}{3}} = \frac{3}{8} = 0.375$.

So, $P[A \cup (A' \cap B)] = P(A) + P(A')P(B) = 0.4 + (0.6)(0.375) = 0.625$.

2.93 Let A and B represent the availability of each fire engine.

(a) $P(A' \cap B') = P(A')P(B') = (0.04)(0.04) = 0.0016$.

(b) $P(A \cup B) = 1 - P(A' \cap B') = 1 - 0.0016 = 0.9984$.

2.95 Consider the events:

A_1: aspirin tablets are selected from the overnight case,

A_2: aspirin tablets are selected from the tote bag,

L_2: laxative tablets are selected from the tote bag,

T_1: thyroid tablets are selected from the overnight case,

T_2: thyroid tablets are selected from the tote bag.

(a) $P(T_1 \cap T_2) = P(T_1)P(T_2) = (3/5)(2/6) = 1/5$.

(b) $P(T_1' \cap T_2') = P(T_1')P(T_2') = (2/5)(4/6) = 4/15$.

(c) $1 - P(A_1 \cap A_2) - P(T_1 \cap T_2) = 1 - P(A_1)P(A_2) - P(T_1)P(T_2) = 1 - (2/5)(3/6) - (3/5)(2/6) = 3/5$.

2.97 (a) $P(Q_1 \cap Q_2 \cap Q_3 \cap Q_4) = P(Q_1)P(Q_2 \mid Q_1)P(Q_3 \mid Q_1 \cap Q_2)P(Q_4 \mid Q_1 \cap Q_2 \cap Q_3) = (15/20)(14/19)(13/18)(12/17) = 91/323$.

(b) Let A be the event that 4 good quarts of milk are selected. Then

$$P(A) = \frac{\binom{15}{4}}{\binom{20}{4}} = \frac{91}{323}.$$

2.99 This is a parallel system of two series subsystems.

    (a) $P = 1 - [1 - (0.7)(0.7)][1 - (0.8)(0.8)(0.8)] = 0.75112$.

    (b) $P = \frac{P(A' \cap C \cap D \cap E)}{P_{\text{system works}}} = \frac{(0.3)(0.8)(0.8)(0.8)}{0.75112} = 0.2045$.

2.101 Consider the events:

    C: an adult selected has cancer,

    D: the adult is diagnosed as having cancer.

    $P(C) = 0.05$, $P(D \mid C) = 0.78$, $P(C') = 0.95$ and $P(D \mid C') = 0.06$. So, $P(D) = P(C \cap D) + P(C' \cap D) = (0.05)(0.78) + (0.95)(0.06) = 0.096$.

2.103 $P(C \mid D) = \frac{P(C \cap D)}{P(D)} = \frac{0.039}{0.096} = 0.40625$.

2.105 Consider the events:

    A: no expiration date,

    B_1: John is the inspector, $P(B_1) = 0.20$ and $P(A \mid B_1) = 0.005$,

    B_2: Tom is the inspector, $P(B_2) = 0.60$ and $P(A \mid B_2) = 0.010$,

    B_3: Jeff is the inspector, $P(B_3) = 0.15$ and $P(A \mid B_3) = 0.011$,

    B_4: Pat is the inspector, $P(B_4) = 0.05$ and $P(A \mid B_4) = 0.005$,

    $P(B_1 \mid A) = \frac{(0.005)(0.20)}{(0.005)(0.20)+(0.010)(0.60)+(0.011)(0.15)+(0.005)(0.05)} = 0.1124$.

2.107   (a) $P(A \cap B \cap C) = P(C \mid A \cap B)P(B \mid A)P(A) = (0.20)(0.75)(0.3) = 0.045$.

    (b) $P(B' \cap C) = P(A \cap B' \cap C) + P(A' \cap B' \cap C) = P(C \mid A \cap B')P(B' \mid A)P(A) + P(C \mid A' \cap B')P(B' \mid A')P(A') = (0.80)(1-0.75)(0.3) + (0.90)(1-0.20)(1-0.3) = 0.564$.

    (c) Use similar argument as in (a) and (b), $P(C) = P(A \cap B \cap C) + P(A \cap B' \cap C) + P(A' \cap B \cap C) + P(A' \cap B' \cap C) = 0.045 + 0.060 + 0.021 + 0.504 = 0.630$.

    (d) $P(A \mid B' \cap C) = P(A \cap B' \cap C)/P(B' \cap C) = (0.06)(0.564) = 0.1064$.

Chapter 3

Random Variables and Probability Distributions

3.1 Discrete; continuous; continuous; discrete; discrete; continuous.

3.3 A table of sample space and assigned values of the random variable is shown next.

Sample Space	w
HHH	3
HHT	1
HTH	1
THH	1
HTT	-1
THT	-1
TTH	-1
TTT	-3

3.5 (a) $c = 1/30$ since $1 = \sum\limits_{x=0}^{3} c(x^2 + 4) = 30c$.

 (b) $c = 1/10$ since

$$1 = \sum_{x=0}^{2} c \binom{2}{x} \binom{3}{3-x} = c \left[\binom{2}{0}\binom{3}{3} + \binom{2}{1}\binom{3}{2} + \binom{2}{2}\binom{3}{1} \right] = 10c.$$

3.7 (a) $P(X < 1.2) = \int_0^1 x\, dx + \int_1^{1.2} (2 - x)\, dx = \frac{x^2}{2}\Big|_0^1 + \left(2x - \frac{x^2}{2} \right)\Big|_1^{1.2} = 0.68.$

 (b) $P(0.5 < X < 1) = \int_{0.5}^1 x\, dx = \frac{x^2}{2}\Big|_{0.5}^1 = 0.375.$

3.9 (a) $P(0 < X < 1) = \int_0^1 \frac{2(x+2)}{5}\, dx = \frac{(x+2)^2}{5}\Big|_0^1 = 1.$

(b) $P(1/4 < X < 1/2) = \int_{1/4}^{1/2} \frac{2(x+2)}{5}\, dx = \frac{(x+2)^2}{5}\Big|_{1/4}^{1/2} = 19/80.$

3.11 We can select x defective sets from 2, and $3 - x$ good sets from 5 in $\binom{2}{x}\binom{5}{3-x}$ ways. A random selection of 3 from 7 sets can be made in $\binom{7}{3}$ ways. Therefore,

$$f(x) = \frac{\binom{2}{x}\binom{5}{3-x}}{\binom{7}{3}}, \qquad x = 0, 1, 2.$$

In tabular form

x	0	1	2
$f(x)$	2/7	4/7	1/7

The following is a probability histogram:

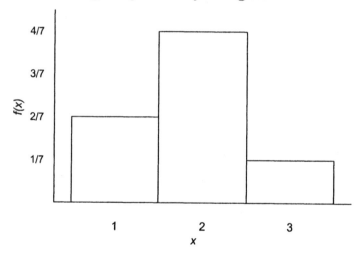

3.13 The c.d.f. of X is

$$F(x) = \begin{cases} 0, & \text{for } x < 0, \\ 0.41, & \text{for } 0 \le x < 1, \\ 0.78, & \text{for } 1 \le x < 2, \\ 0.94, & \text{for } 2 \le x < 3, \\ 0.99, & \text{for } 3 \le x < 4, \\ 1, & \text{for } x \ge 4. \end{cases}$$

3.15 The c.d.f. of X is

$$F(x) = \begin{cases} 0, & \text{for } x < 0, \\ 2/7, & \text{for } 0 \le x < 1, \\ 6/7, & \text{for } 1 \le x < 2, \\ 1, & \text{for } x \ge 2. \end{cases}$$

(a) $P(X = 1) = P(X \le 1) - P(X \le 0) = 6/7 - 2/7 = 4/7$;

(b) $P(0 < X \le 2) = P(X \le 2) - P(X \le 0) = 1 - 2/7 = 5/7$.

3.17 (a) Area $= \int_1^3 (1/2)\, dx = \frac{x}{2}\big|_1^3 = 1$.

(b) $P(2 < X < 2.5) \int_2^{2.5}(1/2)\, dx = \frac{x}{2}\big|_2^{2.5} = \frac{1}{4}$.

(c) $P(X \le 1.6) = \int_1^{1.6}(1/2)\, dx = \frac{x}{2}\big|_1^{1.6} = 0.3$.

3.19 $F(x) = \int_1^x (1/2)\, dt = \frac{x-1}{2}$,
$P(2 < X < 2.5) = F(2.5) - F(2) = \frac{1.5}{2} - \frac{1}{2} = \frac{1}{4}$.

3.21 (a) $1 = k \int_0^1 \sqrt{x}\, dx = \frac{2k}{3}x^{3/2}\big|_0^1 = \frac{2k}{3}$. Therefore, $k - \frac{3}{2}$.

(b) $F(x) = \frac{3}{2}\int_0^x \sqrt{t}\, dt = t^{3/2}\big|_0^x = x^{3/2}$.
$P(0.3 < X < 0.6) = F(0.6) - F(0.3) = (0.6)^{3/2} - (0.3)^{3/2} = 0.3004$.

3.23 The c.d.f. of X is

$$
F(x) = \begin{cases}
0, & \text{for } w < -3, \\
1/27, & \text{for } -3 \le w < -1, \\
7/27, & \text{for } -1 \le w < 1, \\
19/27, & \text{for } 1 < w < 3, \\
1, & \text{for } w \ge 3,
\end{cases}
$$

(a) $P(W > 0 = 1 - P(W \le 0) = 1 - 7/27 = 20/27$.

(b) $P(-1 \le W < 3) = F(2) - F(-3) = 19/27 - 1/27 = 2/3$.

3.25 Let T be the total value of the three coins. Let D and N stand for a dime and nickel, respectively. Since we are selecting without replacement, the sample space containing elements for which $t = 20, 25,$ and 30 cents corresponding to the selecting of 2 nickels and 1 dime, 1 nickel and 2 dimes, and 3 dimes. Therefore, $P(T = 20) = \frac{\binom{2}{2}\binom{4}{1}}{\binom{6}{3}} = \frac{1}{5}$,

$P(T = 25) = \frac{\binom{2}{1}\binom{4}{2}}{\binom{6}{3}} = \frac{3}{5}$,

$P(T = 30) = \frac{\binom{4}{3}}{\binom{6}{3}} = \frac{1}{5}$,

and the probability distribution in tabular form is

t	20	25	30
$P(T = t)$	1/5	3/5	1/5

As a probability histogram

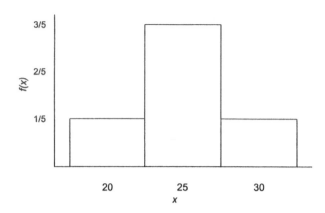

3.27 (a) For $x \geq 0$, $F(x) = \int_0^x \frac{1}{2000} \exp(-t/2000) \, dt = -\exp(-t/2000)|_0^x$
$= 1 - \exp(-x/2000)$. So

$$F(x) = \begin{cases} 0, & x < 0, \\ 1 - \exp(-x/2000), & x \geq 0. \end{cases}$$

(b) $P(X > 1000) = 1 - F(1000) = 1 - [1 - \exp(-1000/2000)] = 0.6065$.

(c) $P(X < 2000) = F(2000) = 1 - \exp(-2000/2000) = 0.6321$.

3.29 (a) $f(x) \geq 0$ and $\int_1^\infty 3x^{-4} \, dx = -3 \left. \frac{x^{-3}}{3} \right|_1^\infty = 1$. So, this is a density function.

(b) For $x \geq 1$, $F(x) = \int_1^x 3t^{-4} \, dt = 1 - x^{-3}$. So,

$$F(x) = \begin{cases} 0, & x < 1, \\ 1 - x^{-3}, & x \geq 1. \end{cases}$$

(c) $P(X > 4) = 1 - F(4) = 4^{-3} = 0.0156$.

3.31 (a) For $y \geq 0$, $F(y) = \frac{1}{4} \int_0^y e^{-t/4} \, dy = 1 - e^{y/4}$. So, $P(Y > 6) = e^{-6/4} = 0.2231$. This probability certainly cannot be considered as "unlikely."

(b) $P(Y \leq 1) = 1 - e^{-1/4} = 0.2212$, which is not so small either.

3.33 (a) Using integral by parts and setting $1 = k \int_0^1 y^4 (1 - y)^3 \, dy$, we obtain $k = 280$.

(b) For $0 \leq y < 1$, $F(y) = 56y^5(1 - Y)^3 + 28y^6(1 - y)^2 + 8y^7(1 - y) + y^8$. So, $P(Y \leq 0.5) = 0.3633$.

(c) Using the cdf in (b), $P(Y > 0.8) = 0.0563$.

3.35 (a) $P(X > 8) = 1 - P(X \leq 8) = \sum_{x=0}^{8} e^{-6} \frac{6^x}{x!} = e^{-6} \left(\frac{6^0}{0!} + \frac{6^1}{1!} + \cdots + \frac{6^8}{8!} \right) = 0.1528$.

(b) $P(X = 2) = e^{-6} \frac{6^2}{2!} = 0.0446$.

3.37 (a) $\sum_{x=0}^{3}\sum_{y=0}^{3} f(x,y) = c\sum_{x=0}^{3}\sum_{y=0}^{3} xy = 36c = 1.$ Hence $c = 1/36.$

(b) $\sum_{x}\sum_{y} f(x,y) = c\sum_{x}\sum_{y} |x - y| = 15c = 1.$ Hence $c = 1/15.$

3.39 (a) We can select x oranges from 3, y apples from 2, and $4 - x - y$ bananas from 3 in $\binom{3}{x}\binom{2}{y}\binom{3}{4-x-y}$ ways. A random selection of 4 pieces of fruit can be made in $\binom{8}{4}$ ways. Therefore,

$$f(x,y) = \frac{\binom{3}{x}\binom{2}{y}\binom{3}{4-x-y}}{\binom{8}{4}}, \qquad x = 0,1,2,3; \quad y = 0,1,2; \quad 1 \le x + y \le 4.$$

(b) $P[(X,Y) \in A] = P(X + Y \le 2) = f(1,0) + f(2,0) + f(0,1) + f(1,1) + f(0,2)$
$= 3/70 + 9/70 + 2/70 + 18/70 + 3/70 = 1/2.$

3.41 (a) $P(X + Y \le 1/2) = \int_0^{1/2}\int_0^{1/2-y} 24xy \, dx \, dy = 12\int_0^{1/2} \left(\frac{1}{2} - y\right)^2 y \, dy = \frac{1}{16}.$

(b) $g(x) = \int_0^{1-x} 24xy \, dy = 12x(1-x)^2,$ for $0 \le x < 1.$

(c) $f(y|x) = \frac{24xy}{12x(1-x)^2} - \frac{2y}{(1-x)^2},$ for $0 \le y \le 1 - x.$
Therefore, $P(Y < 1/8 \mid X = 3/4) = 32\int_0^{1/8} y \, dy = 1/4.$

3.43 (a) $P(0 \le X \le 1/2, \; 1/4 \le Y \le 1/2) = \int_0^{1/2}\int_{1/4}^{1/2} 4xy \, dy \, dx = 3/8\int_0^{1/2} x \, dx = 3/64.$

(b) $P(X < Y) = \int_0^1\int_0^y 4xy \, dx \, dy - 2\int_0^1 y^3 \, dy = 1/2.$

3.45 $P(X + Y > 1/2) = 1 - P(X + Y < 1/2) = 1 - \int_0^{1/4}\int_x^{1/2-x} \frac{1}{y} \, dy \, dx$
$= 1 - \int_0^{1/4} \left[\ln\left(\frac{1}{2} - x\right) - \ln x\right] dx = 1 + \left[\left(\frac{1}{2} - x\right)\ln\left(\frac{1}{2} - x\right) - x\ln x\right]\Big|_0^{1/4}$
$= 1 + \frac{1}{4}\ln\left(\frac{1}{4}\right) = 0.6534.$

3.47 (a) $g(x) = 2\int_x^1 dy = 2(1 - x)$ for $0 < x < 1;$
$h(y) = 2\int_0^y dx = 2y,$ for $0 < y < 1.$
Since $f(x,y) \neq g(x)h(y),$ X and Y are not independent.

(b) $f(x|y) = f(x,y)/h(y) = 1/y,$ for $0 < x < y.$
Therefore, $P(1/4 < X < 1/2 \mid Y = 3/4) = \frac{4}{3}\int_{1/4}^{1/2} dx = \frac{1}{3}.$

3.49 (a)

x	1	2	3
$g(x)$	0.10	0.35	0.55

(b)

y	1	2	3
$h(y)$	0.20	0.50	0.30

(c) $P(Y = 3 \mid X = 2) = \frac{0.2}{0.05+0.10+0.20} = 0.5714.$

3.51 (a) Let X be the number of 4's and Y be the number of 5's. The sample space consists of 36 elements each with probability 1/36 of the form (m,n) where

m is the outcome of the first roll of the die and n is the value obtained on the second roll. The joint probability distribution $f(x,y)$ is defined for $x = 0, 1, 2$ and $y = 0, 1, 2$ with $0 \le x + y \le 2$. To find $f(0,1)$, for example, consider the event A of obtaining zero 4's and one 5 in the 2 rolls. Then $A = \{(1,5),(2,5),(3,5),(6,5),(5,1),(5,2),(5,3),(5,6)\}$, so $f(0,1) = 8/36 = 2/9$. In a like manner we find $f(0,0) = 16/36 = 4/9$, $f(0,2) = 1/36$, $f(1,0) = 2/9$, $f(2,0) = 1/36$, and $f(1,1) = 1/18$.

(b) $P[(X,Y) \in A] = P(2X + Y < 3) = f(0,0) + f(0,1) + f(0,2) + f(1,0) = 4/9 + 1/9 + 1/36 + 2/9 = 11/12$.

3.53 (a) If (x,y) represents the selection of x kings and y jacks in 3 draws, we must have $x = 0,1,2,3$; $y = 0,1,2,3$; and $0 \le x + y \le 3$. Therefore, $(1,2)$ represents the selection of 1 king and 2 jacks which will occur with probability

$$f(1,2) = \frac{\binom{4}{1}\binom{4}{2}}{\binom{12}{3}} = \frac{6}{55}.$$

Proceeding in a similar fashion for the other possibilities, we arrive at the following joint probability distribution:

	$f(x,y)$	0	1	2	3
	0	1/55	6/55	6/55	1/55
y	1	6/55	16/55	6/55	
	2	6/55	6/55		
	3	1/55			

(with x labeling the columns)

(b) $P[(X,Y) \in A] = P(X + Y \ge 2) = 1 - P(X + Y < 2) = 1 - 1/55 - 6/55 - 6/55 = 42/55$.

3.55 $g(x) = \frac{1}{8}\int_2^4 (6 - x - y)\, dy = \frac{3-x}{4}$, for $0 < x < 2$.
So, $f(y|x) = \frac{f(x,y)}{g(x)} = \frac{6-x-y}{2(3-x)}$, for $2 < y < 4$,
and $P(1 < Y < 3 \mid X = 1) = \frac{1}{4}\int_2^3 (5 - y)\, dy = \frac{5}{8}$.

3.57 X and Y are independent since $f(x,y) = g(x)h(y)$ for all (x,y).

3.59 (a) $1 = k\int_0^1 \int_0^1 \int_0^2 xy^2 z\, dx\, dy\, dz = 2k\int_0^1 \int_0^1 y^2 z\, dy\, dz = \frac{2k}{3}\int_0^1 z\, dz = \frac{k}{3}$. So, $k = 3$.

(b) $P\left(X < \frac{1}{4}, Y > \frac{1}{2}, 1 < Z < 2\right) = 3\int_0^{1/4}\int_{1/2}^1 \int_1^2 xy^2 z\, dx\, dy\, dz = \frac{9}{2}\int_0^{1/4}\int_{1/2}^1 y^2 z\, dy\, dz$
$= \frac{21}{16}\int_0^{1/4} z\, dz = \frac{21}{512}$.

3.61 $g(x) = k\int_{30}^{50}(x^2 + y^2)\, dy = k\left(x^2 y + \frac{y^3}{3}\right)\Big|_{30}^{50} = k\left(20x^2 + \frac{98,000}{3}\right)$, and
$h(y) = k\left(20y^2 + \frac{98,000}{3}\right)$.
Since $f(x,y) \ne g(x)h(y)$, X and Y are not independent.

Chapter 4

Mathematical Expectation

4.1 $E(X) = \frac{1}{\pi a^2} \int_{-a}^{a} \int_{-\sqrt{a^2-y^2}}^{\sqrt{a^2-y^2}} x \; dx \; dy = \frac{1}{\pi a^2} \left[\left(\frac{a^2-y^2}{2} \right) - \left(\frac{a^2-y^2}{2} \right) \right] \; dy = 0.$

4.3 $\mu = E(X) = (20)(1/5) + (25)(3/5) + (30)(1/5) = 25$ cents.

4.5 $\mu = E(X) = (0)(0.41) + (1)(0.37) + (2)(0.16) + (3)(0.05) + (4)(0.01) = 0.88.$

4.7 Expected gain $= E(X) = (4000)(0.3) + (-1000)(0.7) = \$500.$

4.9 Let $c =$ amount to play the game and $Y =$ amount won.

y	$5 - c$	$3 - c$	$-c$
$f(y)$	2/13	2/13	9/13

$E(Y) = (5 - c)(2/13) + (3 - c)(2/13) + (-c)(9/13) = 0.$ So, $13c = 16$ which implies $c = \$1.23.$

4.11 For the insurance of \$200,000 pilot, the distribution of the claim the insurance company would have is as follows:

Claim Amount	\$200,000	\$100,000	\$50,000	0
$f(x)$	0.002	0.01	0.1	0.888

So, the expected claim would be

$(\$200,000)(0.002) + (\$100,000)(0.01) + (\$50,000)(0.1) + (\$0)(0.888) = \$6,400.$

Hence the insurance company should charge a premium of $\$6,400 + \$500 = \$6,900.$

4.13 $E(X) = \frac{4}{\pi} \int_0^1 \frac{x}{1+x^2} \; dx = \frac{\ln 4}{\pi}.$

4.15 $E(X) = \int_0^1 x^2 \; dx + \int_1^2 x(2 - x) \; dx = 1.$ Therefore, the average number of hours per year is $(1)(100) = 100$ hours.

4.17 The probability density function is,

x	-3	6	9
$f(x)$	1/6	1/2	1/3
$g(x)$	25	169	361

$$\mu_{g(X)} = E[(2X+1)^2] = (25)(1/6) + (169)(1/2) + (361)(1/3) = 209.$$

4.19 Let $Y = 1200X - 50X^2$ be the amount spent.

x	0	1	2	3
$f(x)$	1/10	3/10	2/5	1/5
$y = g(x)$	0	1150	2200	3150

$$\mu_Y = E(1200X - 50X^2) = (0)(1/10) + (1150)(3/10) + (2200)(2/5) + (3150)(1/5)$$
$$= \$1,855.$$

4.21 $E(X^2) = \int_0^1 2x^2(1-x)\,dx = \frac{1}{6}$. Therefore, the average profit per new automobile is $(1/6)(\$5000.00) = \$833.33$.

4.23 (a) $E[g(X,Y)] = E(XY^2) = \sum_x \sum_y xy^2 f(x,y)$
$$= (2)(1)^2(0.10) + (2)(3)^2(0.20) + (2)(5)^2(0.10) + (4)(1)^2(0.15) + (4)(3)^2(0.30)$$
$$+ (4)(5)^2(0.15) = 35.2.$$

(b) $\mu_X = E(X) = (2)(0.40) + (4)(0.60) = 3.20$,
$\mu_Y = E(Y) = (1)(0.25) + (3)(0.50) + (5)(0.25) = 3.00.$

4.25 $\mu_{X+Y} = E(X+Y) = \sum_{x=0}^{3} \sum_{y=0}^{3} (x+y)f(x,y) = (0+0)(1/55) + (1+0)(6/55) + \cdots + (0+3)(1/55) = 2.$

4.27 $E(X) = \frac{1}{2000} \int_0^{\infty} x \exp(-x/2000)\,dx = 2000 \int_0^{\infty} y \exp(-y)\,dy = 2000.$

4.29 (a) The density function is shown next

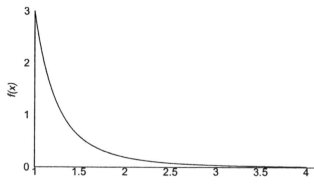

(b) $\mu = E(X) = \int_1^{\infty} 3x^{-3}\,dx = \frac{3}{2}.$

4.31 (a) $\mu = E(Y) = 5\int_0^1 y(1-y)^4\,dy = -\int_0^1 y\,d(1-y)^5 = \int_0^\infty (1-y)^5\,dy = \frac{1}{6}$.

(b) $P(Y > 1/6) = \int_{1/6}^1 5(1-y)^4\,dy = --(1-y)^5|_{1/6}^1 = (1-1/6)^5 = 0.4019$.

4.33 $\mu = \$500$. So, $\sigma^2 = E[(X-\mu)^2] = \sum_x (x-\mu)^2 f(x) = (-1500)^2(0.7) + (3500)^2(0.3) = \$5,250,000$.

4.35 $\mu = (2)(0.01) + (3)(0.25) + (4)(0.4) + (5)(0.3) + (6)(0.04) = 4.11$,
$E(X^2) = (2)^2(0.01) + (3)^2(0.25) + (4)^2(0.4) + (5)^2(0.3) + (6)^2(0.04) = 17.63$.
So, $\sigma^2 = 17.63 - 4.11^2 = 0.74$.

4.37 It is know $\mu = 1/3$.
So, $E(X^2) = \int_0^1 2x^2(1-x)\,dx = 1/6$ and $\sigma^2 = 1/6 - (1/3)^2 = 1/18$. So, in the actual profit, the variance is $\frac{1}{18}(5000)^2$.

4.39 It is known $\mu = 1$.
Since $E(X^2) = \int_0^1 x^2\,dx + \int_1^2 x^2(2-x)\,dx = 7/6$, then $\sigma^2 = 7/6 - (1)^2 = 1/6$.

4.41 It is known $\mu_{g(X)} = E[(2X+1)^2] = 209$. Hence
$\sigma_{g(X)}^2 = \sum_x [(2X+1)^2 - 209]^2 g(x)$
$= (25-209)^2(1/6) + (169-209)^2(1/2) + (361-209)^2(1/3) = 14,144$.
So, $\sigma_{g(X)} = \sqrt{14,144} = 118.9$.

4.43 $\mu_Y = E(3X-2) = \frac{1}{4}\int_0^\infty (3x-2)e^{-x/4}\,dx = 10$. So
$\sigma_Y^2 = E\{[(3X-2) - 10]^2\} = \frac{9}{4}\int_0^\infty (x-4)^2 e^{-x/4}\,dx = 144$.

4.45 $\mu_X = \sum_x xg(x) = 2.45$, $\mu_Y = \sum_y yh(y) = 2.10$, and
$E(XY) = \sum_x \sum_x xyf(x,y) = (1)(0.05) + (2)(0.05) + (3)(0.10) + (2)(0.05)$
$+ (4)(0.10) + (6)(0.35) + (3)(0) + (6)(0.20) + (9)(0.10) = 5.15$.
So, $\sigma_{XY} = 5.15 - (2.45)(2.10) = 0.005$.

4.47 $g(x) = \frac{2}{3}\int_0^1 (x+2y)\,dy = \frac{2}{3}(x+1)$, for $0 < x < 1$, so $\mu_X = \frac{2}{3}\int_0^1 x(x+1)\,dx = \frac{5}{9}$;
$h(y) = \frac{2}{3}\int_0^1 (x+2y)\,dx = \frac{2}{3}\left(\frac{1}{2} + 2y\right)$, so $\mu_Y = \frac{2}{3}\int_0^1 y\left(\frac{1}{2} + 2y\right)\,dy = \frac{11}{18}$; and
$E(XY) = \frac{2}{3}\int_0^1 \int_0^1 xy(x+2y)\,dy\,dx = \frac{1}{3}$.
So, $\sigma_{XY} = E(XY) - \mu_X\mu_Y = \frac{1}{3} - \left(\frac{5}{9}\right)\left(\frac{11}{18}\right) = -0.0062$.

4.49 $E(X) = (0)(0.41) + (1)(0.37) + (2)(0.16) + (3)(0.05) + (4)(0.01) = 0.88$
and $E(X^2) = (0)^2(0.41) + (1)^2(0.37) + (2)^2(0.16) + (3)^2(0.05) + (4)^2(0.01) = 1.62$.
So, $Var(X) = 1.62 - 0.88^2 = 0.8456$ and $\sigma = \sqrt{0.8456} = 0.9196$.

4.51 Previously we found $\mu = 4.11$ and $\sigma^2 = 0.74$, Therefore,
$\mu_{g(X)} = E(3X-2) = 3\mu - 2 = (3)(4.11) - 2 = 10.33$ and $\sigma_{g(X)} = 9\sigma^2 = 6.66$.

4.53 Let X = number of cartons sold and Y = profit.
We can write $Y = 1.65X + (0.90)(5 - X) - 6 = 0.75X - 1.50$. Now
$E(X) = (0)(1/15) + (1)(2/15) + (2)(2/15) + (3)(3/15) + (4)(4/15) + (5)(3/15) = 46/15$,
and $E(Y) = (0.75)E(X) - 1.50 = (0.75)(46/15) - 1.50 = \$0.80$.

4.55 $E(X) = (-3)(1/6) + (6)(1/2) + (9)(1/3) = 11/2$,
$E(X^2) = (-3)^2(1/6) + (6)^2(1/2) + (9)^2(1/3) = 93/2$. So,
$E[(2X + 1)^2] = 4E(X^2) + 4E(X) + 1 = (4)(93/2) + (4)(11/2) + 1 = 209$.

4.57 The equations $E[(X - 1)^2] = 10$ and $E[(X - 2)^2] = 6$ may be written in the form:

$$E(X^2) - 2E(X) = 9, \qquad E(X^2) - 4E(X) = 2.$$

Solving these two equations simultaneously we obtain

$$E(X) = 7/2, \quad \text{and} \quad E(X^2) = 16.$$

Hence $\mu = 7/2$ and $\sigma^2 = 16 - (7/2)^2 = 15/4$.

4.59 $E(2XY^2 - X^2Y) = 2E(XY^2) - E(X^2Y)$. Now,
$E(XY^2) = \sum_{x=0}^{2} \sum_{y=0}^{2} xy^2 f(x, y) = (1)(1)^2(3/14) = 3.14$, and
$E(X^2Y) = \sum_{x=0}^{2} \sum_{y=0}^{2} x^2y f(x, y) = (1)^2(1)(3/14) = 3.14$.
Therefore, $E(2XY^2 - X^2Y) = (2)(3/14) - (3/14) = 3/14$.

4.61 $\mu = 900$ hours and $\sigma = 50$ hours. Solving $\mu - k\sigma = 700$ we obtain $k = 4$.
So, using Chebyshev's theorem with $P(\mu - 4\sigma < X < \mu + 4\sigma) \geq 1 - 1/4^2 = 0.9375$,
we obtain $P(700 < X < 1100) \geq 0.9375$. Therefore, $P(X \leq 700) \leq 0.03125$.

4.63 $n = 500$, $\mu = 4.5$ and $\sigma = 2.8733$. Solving $\mu + k(\sigma/\sqrt{500}) = 5$ we obtain

$$k = \frac{5 - 4.5}{2.87333/\sqrt{500}} = \frac{0.5}{0.1284} = 3.8924.$$

So, $P(4 \leq \bar{X} \leq 5) \geq 1 - \frac{1}{k^2} = 0.9340$.

4.65 $\sigma_Z^2 = \sigma_{-2X+4Y-3}^2 = 4\sigma_X^2 + 16\sigma_Y^2 - 16\sigma_{XY} = (4)(5) + (16)(3) - (16)(1) = 52$.

4.67 (a) $P(|X - 10| \geq 3) = 1 - P(|X - 10| < 3)$
$= 1 - P[10 - (3/2)(2) < X < 10 + (3/2)(2)] \leq 1 - \left[1 - \frac{1}{(3/2)^2}\right] = \frac{4}{9}$.

(b) $P(|X - 10| < 3) = 1 - P(|X - 10| \geq 3) \geq 1 - \frac{4}{9} = \frac{5}{9}$.

(c) $P(5 < X < 15) = P[10 - (5/2)(2) < X < 10 + (5/2)(2)] \geq 1 - \frac{1}{(5/2)^2} = \frac{21}{25}$.

(d) $P(|X - 10| \geq c) \leq 0.04$ implies that $P(|X - 10| < c) \geq 1 - 0.04 = 0.96$.
Solving $0.96 = 1 - \frac{1}{k^2}$ we obtain $k = 5$. So, $c = k\sigma = (5)(2) = 10$.

4.69 It is easy to see that the expectations of X and Y are both 3.5. So,

    (a) $E(X+Y) = E(X) + E(Y) = 3.5 + 3.5 = 7.0$.

    (b) $E(X-Y) = E(X) - E(Y) = 0$.

    (c) $E(XY) = E(X)E(Y) = (3.5)(3.5) = 12.25$.

4.71 $E[g(X,Y)] = E(X/Y^3 + X^2Y) = E(X/Y^3) + E(X^2Y)$.

$E(X/Y^3) = \int_1^2 \int_0^1 \frac{2x(x+2y)}{7y^3}\, dx\, dy = \frac{2}{7}\int_1^2 \left(\frac{1}{3y^3} + \frac{1}{y^2}\right)\, dy = \frac{15}{84}$;

$E(X^2Y) = \int_1^2 \int_0^1 \frac{2x^2y(x+2y)}{7}\, dx\, dy = \frac{2}{7}\int_1^2 y\left(\frac{1}{4} + \frac{2y}{3}\right)\, dy = \frac{139}{252}$.

Hence, $E[g(X,Y)] = \frac{15}{84} + \frac{139}{252} = \frac{46}{63}$.

4.73   (a) $\mu = \frac{1}{5}\int_0^5 x\, dx = 2.5$, $\sigma^2 = E(X^2) - \mu^2 = \frac{1}{5}x^2 \int_0^5 x^2\, dx - 2.5^2 = 2.08$.

      So, $\sigma = \sqrt{\sigma^2} = 1.44$.

    (b) By Chebyshev's theorem,

$$P[2.5 - (2)(1.44) < X < 2.5 + (2)(1.44)] = P(-0.38 < X < 5.38) \geq 0.75.$$

Using integration, $P(-0.38 < X < 5.38) = 1 \geq 0.75$;

$$P[2.5 - (3)(1.44) < X < 2.5 + (3)(1.44)] = P(-1.82 < X < 6.82) \geq 0.89.$$

Using integration, $P(-1.82 < X < 6.82) = 1 \geq 0.89$.

4.75 For $0 < a < 1$, since $g(a) = \sum_{x=0}^{\infty} a^x = \frac{1}{1-a}$, $g'(a) = \sum_{x=1}^{\infty} xa^{x-1} = \frac{1}{(1-a)^2}$ and

$g''(a) = \sum_{x=2}^{\infty} x(x-1)a^{x-2} = \frac{2}{(1-a)^3}$.

    (a) $E(X) = (3/4)\sum_{x=1}^{\infty} x(1/4)^x = (3/4)(1/4)\sum_{x=1}^{\infty} x(1/4)^{x-1} = (3/16)[1/(1-1/4)^2]$

      $= 1/3$, and $E(Y) = E(X) = 1/3$.

      $E(X^2) - E(X) = E[X(X-1)] = (3/4)\sum_{x=2}^{\infty} x(x-1)(1/4)^x$

      $= (3/4)(1/4)^2 \sum_{x=2}^{\infty} x(x-1)(1/4)^{x-2} = (3/4^3)[2/(1-1/4)^3] = 2/9$.

      So, $Var(X) = E(X^2) - [E(X)]^2 = [E(X^2) - E(X)] + E(X) - [E(X)]^2$

      $2/9 + 1/3 - (1/3)^2 = 4/9$, and $Var(Y) = 4/9$.

    (b) $E(Z) = E(X) + E(Y) = (1/3) + (1/3) = 2/3$, and

      $Var(Z) = Var(X+Y) = Var(X) + Var(Y) = (4/9) + (4/9) = 8/9$, since X and

      Y are independent (from Exercise 3.79).

4.77   (a) $E(Y) = \int_0^{\infty} ye^{-y/4}\, dy = 4$.

    (b) $E(Y^2) = \int_0^{\infty} y^2 e^{-y/4}\, dy = 32$ and $Var(Y) = 32 - 4^2 = 16$.

4.79 Using the exact formula, we have

$$E(e^Y) = \int_7^8 e^y \, dy = e^y\big|_7^8 = 1884.32.$$

Using the approximation, since $g(y) = e^y$, so $g''(y) = e^y$. Hence, using the approximation formula,

$$E(e^Y) \approx e^{\mu_Y} + e^{\mu_Y}\frac{\sigma_Y^2}{2} = \left(1 + \frac{1}{24}\right)e^{7.5} = 1883.38.$$

The approximation is very close to the true value.

Chapter 5

Some Discrete Probability Distributions

5.1 This is a uniform distribution: $f(x) = \frac{1}{10}$, for $x = 1, 2, \ldots, 10$.

Therefore $P(X < 4) = \sum_{x=1}^{3} f(x) = \frac{3}{10}$.

5.3 $\mu = \sum_{x=1}^{10} \frac{x}{10} = 5.5$, and $\sigma^2 = \sum_{x=1}^{10} \frac{(x-5.5)^2}{10} = 8.25$.

5.5 We are considering a $b(x; 20, 0.3)$.

(a) $P(X \geq 10) = 1 - P(X \leq 9) = 1 - 0.9520 = 0.0480$.

(b) $P(X \leq 4) = 0.2375$.

(c) $P(X = 5) = 0.1789$. This probability is not very small so this is not a rare event. Therefore, $P = 0.30$ is reasonable.

5.7 $p = 0.7$.

(a) For $n = 10$, $P(X < 5) = P(X \leq 4) = 0.0474$.

(b) For $n = 20$, $P(X < 10) = P(X \leq 9) = 0.0171$.

5.9 For $n = 15$ and $p = 0.25$, we have

(a) $P(3 \leq X \leq 6) = P(X \leq 6) - P(X \leq 2) = 0.9434 - 0.2361 = 0.7073$.

(b) $P(X < 4) = P(X \leq 3) = 0.4613$.

(c) $P(X > 5) = 1 - P(X \leq 5) = 1 - 0.8516 = 0.1484$.

5.11 From Table A.1 with $n = 7$ and $p = 0.9$, we have
$P(X = 5) = P(X \leq 5) - P(X \leq 4) = 0.1497 - 0.0257 = 0.1240$.

5.13 From Table A.1 with $n = 5$ and $p = 0.7$, we have
$P(X \geq 3) = 1 - P(X \leq 2) = 1 - 0.1631 = 0.8369$.

5.15 $p = 0.4$ and $n = 5$.

    (a) $P(X = 0) = 0.0778$.

    (b) $P(X < 2) = P(X \leq 1) = 0.3370$.

    (c) $P(X > 3) = 1 - P(X \leq 3) = 1 - 0.9130 = 0.0870$.

5.17 Since $\mu = np = (5)(0.7) = 3.5$ and $\sigma^2 = npq = (5)(0.7)(0.3) = 1.05$ with $\sigma = 1.025$. Then $\mu \pm 2\sigma = 3.5 \pm (2)(1.025) = 3.5 \pm 2.050$ or from 1.45 to 5.55. Therefore, at least 3/4 of the time when 5 people are selected at random, anywhere from 2 to 5 are of the opinion that tranquilizers do not cure but only cover up the real problem.

5.19 Let X_1 = number of times encountered green light with $P(\text{Green}) = 0.35$, X_2 = number of times encountered yellow light with $P(\text{Yellow}) = 0.05$, and X_3 = number of times encountered red light with $P(\text{Red}) = 0.60$. Then

$$f(x_1, x_2, x_3) = \binom{n}{x_1, x_2, x_3} (0.35)^{x_1} (0.05)^{x_2} (0.60)^{x_3}.$$

5.21 Using the multinomial distribution with required probability is

$$\binom{7}{0, 0, 1, 4, 2} (0.02)(0.82)^4 (0.1)^2 = 0.0095.$$

5.23 Using the multinomial distribution, we have

$$\binom{9}{3, 3, 1, 2} (0.4)^3 (0.2)^3 (0.3)(0.2)^2 = 0.0077.$$

5.25 $n = 20$ and the probability of a defective is $p = 0.10$. So, $P(X \leq 3) = 0.8670$.

5.27 $n = 20$ and $p = 0.90$;

    (a) $P(X = 18) = P(X \leq 18) - P(X \leq 17) = 0.6083 - 0.3231 = 0.2852$.

    (b) $P(X \geq 15) = 1 - P(X \leq 14) = 1 - 0.0113 = 0.9887$

    (c) $P(X \leq 18) = 0.6083$.

5.29 Using the hypergeometric distribution, we get

    (a) $\dfrac{\binom{12}{2}\binom{40}{5}}{\binom{52}{7}} = 0.3246$.

    (b) $1 - \dfrac{\binom{48}{7}}{\binom{52}{7}} = 0.4496$.

5.31 Using the hypergeometric distribution, we get $h(2; 9, 6, 4) = \dfrac{\binom{4}{2}\binom{5}{4}}{\binom{9}{6}} = \dfrac{5}{14}$.

5.33 $h(x; 6, 3, 4) = \frac{\binom{4}{x}\binom{2}{3-x}}{\binom{6}{3}}$, for $x = 1, 2, 3$.

   $P(2 \leq X \leq 3) = h(2; 6, 3, 4) + h(3; 6, 3, 4) = \frac{4}{5}$.

5.35 $P(X \leq 2) = \sum_{x=0}^{2} h(x; 50, 5, 10) = 0.9517$.

5.37 (a) $P(X = 0) = b(0; 3, 3/25) = 0.6815$.

   (b) $P(1 \leq X \leq 3) = \sum_{x=1}^{3} b(x; 3, 1/25) = 0.1153$.

5.39 Since $\mu = (13)(13/52) = 3.25$ and $\sigma^2 = (13)(1/4)(3/4)(39/51) = 1.864$ with $\sigma = 1.365$, at least 75% of the time the number of hearts lay between

$$\mu + 2\sigma - 3.25 \pm (2)(1.365) \text{ or from } 0.52 \text{ to } 5.98.$$

5.41 Using the binomial approximation of the hypergeometric with $p = 0.5$, the probability is $1 - \sum_{x=0}^{2} b(x; 10, 0.5) = 0.9453$.

5.43 Using the binomial approximation of the hypergeometric distribution with 0.7, the probability is $1 - \sum_{x=10}^{13} b(x; 18, 0.7) = 0.6077$.

5.45 (a) The extension of the hypergeometric distribution gives a probability

$$\frac{\binom{2}{1}\binom{3}{1}\binom{5}{1}\binom{2}{1}}{\binom{12}{4}} = \frac{4}{33}.$$

   (b) Using the extension of the hypergeometric distribution, we have

$$\frac{\binom{2}{1}\binom{3}{1}\binom{2}{2}}{\binom{12}{4}} + \frac{\binom{2}{2}\binom{3}{1}\binom{2}{1}}{\binom{12}{4}} + \frac{\binom{2}{1}\binom{3}{2}\binom{2}{1}}{\binom{12}{4}} = \frac{8}{165}.$$

5.47 $h(5; 25, 15, 10) = \frac{\binom{10}{5}\binom{15}{10}}{\binom{25}{15}} = 0.2315$.

5.49 (a) $\frac{\binom{3}{0}\binom{17}{5}}{\binom{20}{5}} = 0.3991$.

   (b) $\frac{\binom{3}{2}\binom{17}{3}}{\binom{20}{5}} = 0.1316$.

5.51 Using the negative binomial distribution, the required probability is

$$b^*(10; 5, 0.3) = \binom{9}{4}(0.3)^5(0.7)^5 = 0.0515.$$

5.53 (a) $P(X > 5) = \sum_{x=6}^{\infty} p(x;5) = 1 - \sum_{x=0}^{5} p(x;5) = 0.3840.$

     (b) $P(X = 0) = p(0;5) = 0.0067.$

5.54 (a) Using the negative binomial distribution, we get

$$b^*(7;3,1/2) = \binom{6}{2}(1/2)^7 = 0.1172.$$

     (b) From the geometric distribution, we have $g(4;1/2) = (1/2)(1/2)^3 = 1/16.$

5.55 The probability that all coins turn up the same is 1/4. Using the geometric distribution with $p = 3/4$ and $q = 1/4$, we have

$$P(X < 4) = \sum_{x=1}^{3} g(x;3/4) = \sum_{x=1}^{3}(3/4)(1/4)^{x-1} = \frac{63}{64}.$$

5.57 Using the geometric distribution

     (a) $P(X = 3) = g(3;0.7) = (0.7)(0.3)^2 = 0.0630.$

     (b) $P(X < 4) = \sum_{x=1}^{3} g(x;0.7) = \sum_{x=1}^{3}(0.7)(0.3)^{x-1} = 0.9730.$

5.59 (a) $P(X \geq 4) = 1 - P(X \leq 3) = 0.1429.$

     (b) $P(X = 0) = p(0;2) = 0.1353.$

5.61 (a) Using the negative binomial distribution, we obtain
$b^*(6;4,0.8) = \binom{5}{3}(0.8)^4(0.2)^2 = 0.1638.$

     (b) From the geometric distribution, we have $g(3;0.8) = (0.8)(0.2)^2 = 0.032.$

5.63 (a) Using the Poisson distribution with $\mu = 5$, we find

$$P(X > 5) = 1 - P(X \leq 5) = 1 - 0.6160 = 0.3840.$$

     (b) Using the binomial distribution with $p = 0.3840$, we get

$$b(3;4,0.384) = \binom{4}{3}(0.3840)^3(0.6160) = 0.1395.$$

     (c) Using the geometric distribution with $p = 0.3840$, we have

$$g(5;0.384) = (0.394)(0.616)^4 = 0.0553.$$

5.65 $\mu = np = (10000)(0.001) = 10$, so

$$P(6 \leq X \leq 8) = P(X \leq 8) - P(X \leq 5) \approx \sum_{x=0}^{8} p(x;10) - \sum_{x=0}^{5} p(x;10) = 0.2657.$$

5.67 (a) $\mu = (2000)(0.002) = 4$ and $\sigma^2 = 4$.

(b) For $k = 2$, we have $\mu \pm 2\sigma = 4 \pm 4$ or from 0 to 8.

5.69 (a) $P(X \le 3|\lambda t = 5) = 0.2650$.

(b) $P(X > 1|\lambda t = 5) = 1 - 0.0404 = 0.9596$.

5.71 (a) $P(X > 10|\lambda t = 14) = 1 - 0.1757 = 0.8243$.

(b) $\lambda t = 14$.

5.73 $\mu = (4000)(0.001) = 4$.

5.75 $\mu = \lambda t = (1.5)(5) = 7.5$ and $P(X = 0|\lambda t = 7.5) = e^{-7.5} = 5.53 \times 10^{-4}$.

5.77 (a) $P(X > 10|\lambda t = 5) = 1 - P(X \le 10|\lambda t = 5) = 1 - 0.9863 = 0.0137$.

(b) $\mu = \lambda t = (5)(3) = 15$, so $P(X > 20|\lambda t = 15) = 1 - P(X \le 20|\lambda = 15) = 1 - 0.9170 = 0.0830$.

5.79 So, Let Y = number of shifts until it fails. Then Y follows a geometric distribution with $p = 0.10$. So,

$$P(Y \le 6) = g(1; 0.1) + g(2; 0.1) + \cdots + g(6; 0.1)$$
$$= (0.1)[1 + (0.9) + (0.9)^2 + \cdots + (0.9)^5] = 0.4686.$$

Chapter 6

Some Continuous Probability Distributions

6.1 (a) Area=0.9236.

   (b) Area=$1 - 0.1867 = 0.8133$.

   (c) Area=$0.2578 - 0.0154 = 0.2424$.

   (d) Area=0.0823.

   (e) Area=$1 - 0.9750 = 0.0250$.

   (f) Area=$0.9591 - 0.3156 = 0.6435$.

6.3 (a) From Table A.3, $k = -1.72$.

   (b) Since $P(Z > k) = 0.2946$, then $P(Z < k) = 0.7054/$ From Table A.3, we find $k = 0.54$.

   (c) The area to the left of $z = -0.93$ is found from Table A.3 to be 0.1762. Therefore, the total area to the left of k is 0.1762+0.7235=0.8997, and hence $k = 1.28$.

6.5 (a) $z = (15 - 18)/2.5 = -1.2$; $P(X < 15) = P(Z < -1.2) = 0.1151$.

   (b) $z = -0.76$, $k = (2.5)(-0.76) + 18 = 16.1$.

   (c) $z = 0.91$, $k = (2.5)(0.91) + 18 = 20.275$.

   (d) $z_1 = (17 - 18)/2.5 = -0.4$, $z_2 = (21 - 18)/2.5 = 1.2$;
   $P(17 < X < 21) = P(-0.4 < Z < 1.2) = 0.8849 - 0.3446 = 0.5403$.

6.7 (a) $z = (32 - 40)/6.3 = -1.27$; $P(X > 32) = P(Z > -1.27) = 1 - 0.1020 = 0.8980$.

   (b) $z = (28 - 40)/6.3 = -1.90$, $P(X < 28) = P(Z < -1.90) = 0.0287$.

   (c) $z_1 = (37 - 40)/6.3 = -0.48$, $z_2 = (49 - 40)/6.3 = 1.43$;
   So, $P(37 < X < 49) = P(-0.48 < Z < 1.43) = 0.9236 - 0.3156 = 0.6080$.

6.9 (a) $z = (224 - 200)/15 = 1.6$. Fraction of the cups containing more than 224 millimeters is $P(Z > 1.6) = 0.0548$.

(b) $z_1 = (191 - 200)/15 = -0.6$, $Z_2 = (209 - 200)/15 = 0.6$;
$P(191 < X < 209) = P(-0.6 < Z < 0.6) = 0.7257 - 0.2743 = 0.4514$.

(c) $z = (230 - 200)/15 = 2.0$; $P(X > 230) = P(Z > 2.0) = 0.0228$. Therefore, $(1000)(0.0228) = 22.8$ or approximately 23 cups will overflow.

(d) $z = -0.67$, $x = (15)(-0.67) + 200 = 189.95$ millimeters.

6.11 (a) $z = (30 - 24)/3.8 = 1.58$; $P(X > 30) = P(Z > 1.58) = 0.0571$.

(b) $z = (15 - 24)/3.8 = -2.37$; $P(X > 15) = P(Z > -2.37) = 0.9911$. He is late 99.11% of the time.

(c) $z = (25 - 24)/3.8 = 0.26$; $P(X > 25) = P(Z > 0.26) = 0.3974$.

(d) $z = 1.04$, $x = (3.8)(1.04) + 24 = 27.952$ minutes.

(e) Using the binomial distribution with $p = 0.0571$, we get
$b(2; 3, 0.0571) = \binom{3}{2}(0.0571)^2(0.9429) = 0.0092$.

6.13 $z = -1.88$, $x = (2)(-1.88) + 10 = 6.24$ years.

6.15 $\mu = \$15.90$ and $\sigma = \$1.50$.

(a) 51%, since $P(13.75 < X < 16.22) = P\left(\frac{13.745 - 15.9}{1.5} < Z < \frac{16.225 - 15.9}{1.5}\right)$
$= P(-1.437 < Z < 0.217) = 0.5871 - 0.0749 = 0.5122$.

(b) $\$18.36$, since $P(Z > 1.645) = 0.05$; $x = (1.645)(1.50) + 15.90 + 0.005 = 18.37$.

6.17 (a) $z = (10,175 - 10,000)/100 = 1.75$. Proportion of components exceeding 10.150 kilograms in tensile strength= $P(X > 10,175) = P(Z > 1.75) = 0.0401$.

(b) $z_1 = (9,775 - 10,000)/100 = -2.25$ and $z_2 = (10,225 - 10,000)/100 = 2.25$. Proportion of components scrapped= $P(X < 9,775) + P(X > 10,225) = P(Z < -2.25) + P(Z > 2.25) = 2P(Z < -2.25) = 0.0244$.

6.19 $z = (94.5 - 115)/12 = -1.71$; $P(X < 94.5) = P(Z < -1.71) = 0.0436$. Therefore, $(0.0436)(600) = 26$ students will be rejected.

6.21 $A = 7$ and $B = 10$.

(a) $P(X \leq 8.8) = \frac{8.8 - 7}{3} = 0.60$.

(b) $P(7.4 < X < 9.5) = \frac{9.5 - 7.4}{3} = 0.70$.

(c) $P(X \geq 8.5) = \frac{10 - 8.5}{3} = 0.50$.

6.23 (a) From Table A.1 with $n = 15$ and $p = 0.2$ we have
$$P(1 \leq X \leq 4) = \sum_{x=0}^{4} b(x; 15, 0.2) - b(0; 15, 0.2) = 0.8358 - 0.0352 = 0.8006.$$

(b) By the normal-curve approximation we first find
$\mu = np = 3$ and then $\sigma^2 = npq = (15)(0.2)(0.8) = 2.4$. Then $\sigma = 1.549$.
Now, $z_1 = (0.5 - 3)/1.549 = -1.61$ and $z_2 = (4.5 - 3)/1.549 = 0.97$.
Therefore, $P(1 \leq X \leq 4) = P(-1.61 \leq Z \leq 0.97) = 0.8340 - 0.0537 = 0.7803$.

6.25 $n = 100$.

    (a) $p = 0.01$ with $\mu = (100)(0.01) = 1$ and $\sigma = \sqrt{(100)(0.01)(0.99)} = 0.995$.
        So, $z = (0.5 - 1)/0.995 = -0.503$. $P(X \leq 0) \approx P(Z \leq -0.503) = 0.3085$.

    (b) $p = 0.05$ with $\mu = (100)(0.05) = 5$ and $\sigma = \sqrt{(100)(0.05)(0.95)} = 2.1794$.
        So, $z = (0.5 - 5)/2.1794 = -2.06$. $P(X \leq 0) \approx P(X \leq -2.06) = 0.0197$.

6.27 $\mu = (100)(0.9) = 90$ and $\sigma = \sqrt{(100)(0.9)(0.1)} = 3$.

    (a) $z_1 = (83.5 - 90)/3 = -2.17$ and $z_2 = (95.5 - 90)/3 = 1.83$.
        $P(83.5 < X < 95.5) = P(-2.17 < Z < 1.83) = 0.9664 - 0.0150 = 0.9514$.

    (b) $z = (85.5 - 90)/3 = -1.50$; $P(X < 85.5) = P(Z < -1.50) = 0.0668$.

6.29 $\mu = (1000)(0.2) = 200$ and $\sigma = \sqrt{(1000)(0.2)(0.8)} = 12.649$.

    (a) $z_1 = (169.5 - 200)/12.649 = -2.41$ and $z_2 = (185.5 - 200)/12.649 = -1.15$.
        $P(169.5 < X < 185.5) = P(-2.41 < Z < -1.15) = 0.1251 - 0.0080 = 0.1171$.

    (b) $z_1 = (209.5 - 200)/12.649 = 0.75$ and $z_2 = (225.5 - 200)/12.649 = 2.02$.
        $P(209.5 < X < 225.5) = P(0.75 < Z < 2.02) = 0.9783 - 0.7734 = 0.2049$.

6.31 $\mu = (180)(1/6) = 30$ and $\sigma = \sqrt{(180)(1/6)(5/6)} = 5$ with $z = (35.5 - 30)/5 = 1.1$.
    $P(X > 35.5) = P(Z > 1.1) = 1 - 0.8643 = 0.1357$.

6.33 $\mu = (400)(1/10) = 40$ and $\sigma = \sqrt{(400)(1/10)(9/10)} = 6$.

    (a) $z = (31.5 - 40)/6 = -1.42$; $P(X < 31.5) = P(Z < -1.42) = 0.0778$.

    (b) $z = (49.5 - 40)/6 = 1.58$; $P(X > 49.5) = P(Z > 1.58) = 1 - 0.9429 = 0.0571$.

    (c) $z_1 = (34.5 - 40)/6 = -0.92$ and $z_2 = (46.5 - 40)/6 = 1.08$;
        $P(34.5 < X < 46.5) = P(-0.92 < Z < 1.08) = 0.8599 - 0.1788 = 0.6811$.

6.35 (a) $p = 0.05$, $n = 100$ with $\mu = 5$ and $\sigma = \sqrt{(100)(0.05)(0.95)} = 2.1794$.
        So, $z = (2.5 - 5)/2.1794 = -1.147$; $P(X \geq 2) \approx P(Z \geq -1.147) = 0.8749$.

    (b) $z = (10.5 - 5)/2.1794 = 2.524$; $P(X \geq 10) \approx P(Z > 2.52) = 0.0059$.

6.37 (a) $P(X \geq 230) = P\left(Z > \frac{230 - 170}{30}\right) = 0.0228$.

    (b) Denote by Y the number of students whose serum cholesterol level exceed 230
        among the 300. Then $Y \sim b(y; 300, 0.0228)$ with $\mu = (300)(0.0228) = 6.84$ and
        $\sigma = \sqrt{(300)(0.0228)(1 - 0.0228)} = 2.5854$. So, $z = \frac{8 - 0.5 - 6.84}{2.5854} = 0.26$ and
        $P(X \geq 8) \approx P(Z > 0.26) = 0.3974$.

6.39 $P(1.8 < X < 2.4) = \int_{1.8}^{2.4} xe^{-x}\,dx = \left[-xe^{-x} - e^{-x}\right]\big|_{1.8}^{2.4} = 2.8e^{-1.8} - 3.4e^{-2.4} = 0.1545$.

6.41 Setting $\alpha = 1/2$ in the gamma distribution and integrating, we have

$$\frac{1}{\sqrt{\beta}\Gamma(1/2)} \int_0^\infty x^{-1/2}e^{-x/\beta} \, dx = 1.$$

Substitute $x = y^2/2$, $dx = y \, dy$, to give

$$\Gamma(1/2) = \frac{\sqrt{2}}{\sqrt{\beta}} \int_0^\infty e^{-y^2/2\beta} \, dy = 2\sqrt{\pi} \left(\frac{1}{\sqrt{2\pi}\sqrt{\beta}} \int_0^\infty e^{-y^2/2\beta} \, dy \right) = \sqrt{\pi},$$

since the quantity in parentheses represents one-half of the area under the normal curve $n(y; 0, \sqrt{\beta})$.

6.43 (a) $\mu = \alpha\beta = (2)(3) = 6$ million liters; $\sigma^2 = \alpha\beta^2 = (2)(9) = 18$.

  (b) Water consumption on any given day has a probability of at least 3/4 of falling in the interval $\mu \pm 2\sigma = 6 \pm 2\sqrt{18}$ or from -2.485 to 14.485. That is from 0 to 14.485 million liters.

6.45 $P(X < 3) = \frac{1}{4} \int_0^3 e^{-x/4} \, dx = -e^{-x/4}\big|_0^3 = 1 - e^{-3/4} = 0.5276$.
Let Y be the number of days a person is served in less than 3 minutes. Then

$$P(Y \geq 4) = \sum_{x=4}^{6} b(y; 6, 1 - e^{-3/4}) = \binom{6}{4}(0.5276)^4(0.4724)^2 + \binom{6}{5}(0.5276)^5(0.4724)$$
$$+ \binom{6}{6}(0.5276)^6 = 0.3968.$$

6.47 (a) $E(X) = \int_0^\infty x^2 e^{-x^2/2} \, dx = -xe^{-x^2/2}\big|_0^\infty + \int_0^\infty e^{-x^2/2} \, dx$

$$= 0 + \sqrt{2\pi} \cdot \frac{1}{\sqrt{2\pi}} \int_0^\infty e^{-x^2/2} \, dx = \frac{\sqrt{2\pi}}{2} = \sqrt{\frac{\pi}{2}} = 1.2533.$$

  (b) $P(X > 2) = \int_2^\infty x e^{-x^2/2} \, dx = -e^{-x^2/2}\big|_2^\infty = e^{-2} = 0.1353$.

6.49 $R(t) = ce^{-\int 1/\sqrt{t} \, dt} = ce^{-2\sqrt{t}}$. However, $R(0) = 1$ and hence $c = 1$. Now

$$f(t) = Z(t)R(t) = e^{-2\sqrt{t}}/\sqrt{t}, \qquad t > 0,$$

and

$$P(T > 4) = \int_4^\infty e^{-2\sqrt{t}}/\sqrt{t} \, dt = -e^{-2\sqrt{t}}\big|_4^\infty = e^{-4} = 0.0183.$$

6.51 $\alpha = 5$; $\beta = 10$;

  (a) $\alpha\beta = 50$.

  (b) $\sigma^2 = \alpha\beta^2 = 500$; so $\sigma = \sqrt{500} = 22.36$.

  (c) $P(X > 30) = \frac{1}{\beta^\alpha\Gamma(\alpha)} \int_{30}^\infty x^{\alpha-1}e^{-x/\beta} \, dx$. Using the incomplete gamma with $y = x/\beta$, then

$$1 - P(X \leq 30) = 1 - P(Y \leq 3) = 1 - \int_0^3 \frac{y^4 e^{-y}}{\Gamma(5)} \, dy = 1 - 0.185 = 0.815.$$

6.53 $\mu = 3$ seconds with $f(x) = \frac{1}{3}e^{-x/3}$ for $x > 0$.

    (a) $P(X > 5) = \int_5^\infty \frac{1}{3}e^{-x/3}\,dx = \frac{1}{3}\left[-3e^{-x/3}\right]\big|_5^\infty = e^{-5/3} = 0.1889$.

    (b) $P(X > 10) = e^{-10/3} = 0.0357$.

6.55 $\mu = E(X) = e^{4+4/2} = e^6$; $\sigma^2 = e^{8+4}(e^4 - 1) = e^{12}(e^4 - 1)$.

6.57 (a) $P(X > 1) = 1 - P(X \le 1) = 1 - 10\int_0^1 e^{-10x}\,dx = e^{-10} = 0.000045$.

    (b) $\mu = \beta = 1/10 = 0.1$.

Chapter 7

Functions of Random Variables

7.1 From $y = 2x - 1$ we obtain $x = (y+1)/2$, and given $x = 1, 2,$ and 3, then

$$g(y) = f[(y+1)/2] = 1/3, \qquad \text{for } y = 1, 3, 5.$$

7.3 The inverse functions of $y_1 = x_1 + x_2$ and $y_2 = x_1 - x_2$ are $x_1 = (y_1 + y_2)/2$ and $x_2 = (y_1 - y_2)/2$. Therefore,

$$g(y_1, y_2) = \binom{2}{\frac{y_1+y_2}{2}, \frac{y_1-y_2}{2}, 2-y_1} \left(\frac{1}{4}\right)^{(y_1+y_2)/2} \left(\frac{1}{3}\right)^{(y_1-y_2)/2} \left(\frac{5}{12}\right)^{2-y_1},$$

where $y_1 = 0, 1, 2$, $y_2 = -2, -1, 0, 1, 2$, $y_2 \le y_1$ and $y_1 + y_2 = 0, 2, 4$.

7.5 The inverse function of $y = -2\ln x$ is given by $x = e^{-y/2}$ from which we obtain $|J| = |-e^{-y/2}/2| = e^{-y/2}/2$. Now,

$$g(y) = f(e^{y/2})|J| = e^{-y/2}/2, \qquad y > 0,$$

which is a chi-squared distribution with 2 degrees of freedom.

7.7 To find k we solve the equation $k \int_0^\infty v^2 e^{-bv^2} \, dv = 1$. Let $x = bv^2$, then $dx = 2bv \, dv$ and $dv = \frac{x^{-1/2}}{2\sqrt{b}} \, dx$. Then the equation becomes

$$\frac{k}{2b^{3/2}} \int_0^\infty x^{3/2-1} e^{-x} \, dx = 1, \quad \text{or} \quad \frac{k\Gamma(3/2)}{2b^{3/2}} = 1.$$

Hence $k = \frac{4b^{3/2}}{\Gamma(1/2)}$.

Now the inverse function of $w = mv^2/2$ is $v = \sqrt{2w/m}$, for $w > 0$, from which we obtain $|J| = 1/\sqrt{2mw}$. It follows that

$$g(w) = f(\sqrt{2w/m})|J| = \frac{4b^{3/2}}{\Gamma(1/2)}(2w/m)e^{-2bw/m} = \frac{1}{(m/2b)^{3/2}\Gamma(3/2)}w^{3/2-1}e^{-(2b/m)w},$$

for $w > 0$, which is a gamma distribution with $\alpha = 3/2$ and $\beta = m/2b$.

7.9   (a) The inverse of $y = x + 4$ is $x = y - 4$, for $y > 4$, from which we obtain $|J| = 1$. Therefore,

$$g(y) = f(y-4)|J| = 32/y^3, \quad y > 4.$$

(b) $P(Y > 8) = 32 \int_8^\infty y^{-3} \, dy = -16y^{-2}\big|_8^\infty = \frac{1}{4}.$

7.11 The amount of kerosene left at the end of the day is $Z = Y - X$. Let $W = Y$. The inverse functions of $z = y - x$ and $w = y$ are $x = w - z$ and $y = w$, for $0 < z < w$ and $0 < w < 1$, from which we obtain

$$J = \begin{vmatrix} \frac{\partial x}{\partial w} & \frac{\partial x}{\partial z} \\ \frac{\partial y}{\partial w} & \frac{\partial y}{\partial z} \end{vmatrix} = \begin{vmatrix} 1 & -1 \\ 1 & 0 \end{vmatrix} = 1.$$

Now,

$$g(w, z) = g(w - z, w) = 2, \quad 0 < z < w, \; 0 < w < 1,$$

and the marginal distribution of Z is

$$h(z) = 2\int_z^1 dw = 2(1 - z), \quad 0 < z < 1.$$

7.13 Since I and R are independent, the joint probability distribution is

$$f(i, r) = 12ri(1 - i), \quad 0 < i < 1, \; 0 < r < 1.$$

Let $V = R$. The inverse functions of $w = i^2 r$ and $v = r$ are $i = \sqrt{w/v}$ and $r = v$, for $w < v < 1$ and $0 < w < 1$, from which we obtain

$$J = \begin{vmatrix} \partial i/\partial w & \partial i/\partial v \\ \partial r/\partial w & \partial r/\partial v \end{vmatrix} = \frac{1}{2\sqrt{vw}}.$$

Then,

$$g(w, v) = f(\sqrt{w/v}, v)|J| = 12v\sqrt{w/v}(1 - \sqrt{w/v})\frac{1}{2\sqrt{vw}} = 6(1 - \sqrt{w/v}),$$

for $w < v < 1$ and $0 < w < 1$, and the marginal distribution of W is

$$h(w) = 6\int_w^1 (1 - \sqrt{w/v}) \, dv = 6\,(v - 2\sqrt{wv})\big|_{v=w}^{v=1} = 6 + 6w - 12\sqrt{w}, \quad 0 < w < 1.$$

7.15 The inverse functions of $y = x^2$ are $x_1 = \sqrt{y}$, $x_2 = -\sqrt{y}$ for $0 < y < 1$ and $x_1 = \sqrt{y}$ for $0 < y < 4$. Now $|J_1| = |J_2| = |J_3| = 1/2\sqrt{y}$, from which we get

$$g(y) = f(\sqrt{y})|J_1| + f(-\sqrt{y})|J_2| = \frac{2(\sqrt{y}+1)}{9} \cdot \frac{1}{2\sqrt{y}} + \frac{2(-\sqrt{y}+1)}{9} \cdot \frac{1}{2\sqrt{y}} = \frac{2}{9\sqrt{y}},$$

for $0 < y < 1$ and

$$g(y) = f(\sqrt{y})|J_3| = \frac{2(\sqrt{y}+1)}{9} \cdot \frac{1}{2\sqrt{y}} = \frac{\sqrt{y}+1}{9\sqrt{y}}, \quad \text{for } 1 < y < 4.$$

7.17 The moment-generating function of X is

$$M_X(t) = E(e^{tX}) = \frac{1}{k}\sum_{x=1}^{k} e^{tx} = \frac{e^t(1-e^{kt})}{k(1-e^t)},$$

by summing the geometric series of k terms.

7.19 The moment-generating function of a Poisson random variable is

$$M_X(t) = E(e^{tX}) = \sum_{x=0}^{\infty} \frac{e^{tx}e^{-\mu}\mu^x}{x!} = e^{-\mu}\sum_{x=0}^{\infty} \frac{(\mu e^t)^x}{x!} = e^{-\mu}e^{\mu e^t} - e^{\mu(e^t-1)}.$$

So,

$$\mu = M_X'(0) = \mu\, e^{\mu(e^t-1)+t}\Big|_{t=0} = \mu,$$

$$\mu_2' = M_X''(0) = \mu e^{\mu(e^t-1)+t}(\mu e^t+1)\Big|_{t=0} = \mu(\mu+1),$$

and

$$\sigma^2 = \mu_2' - \mu^2 = \mu(\mu+1) - \mu^2 = \mu.$$

7.21 Using the moment-generating function of the chi-squared distribution, we obtain

$$\mu = M_X'(0) = v(1-2t)^{-v/2-1}\Big|_{t=0} = v,$$
$$\mu_2' = M_X''(0) = v(v+2)\,(1-2t)^{-v/2-2}\Big|_{t=0} = v(v+2).$$

So, $\sigma^2 = \mu_2' - \mu^2 = v(v+2) - v^2 = 2v$.

7.23 The joint distribution of X and Y is $f_{X,Y}(x,y) = e^{-x-y}$ for $x > 0$ and $y > 0$. The inverse functions of $u = x + y$ and $v = x/(x+y)$ are $x = uv$ and $y = u(1-v)$ with

$$J = \begin{vmatrix} v & u \\ 1-v & -u \end{vmatrix} = u \text{ for } u > 0 \text{ and } 0 < v < 1.$$ So, the joint distribution of U and V is

$$f_{U,V}(u,v) = ue^{-uv} \cdot e^{-u(1-v)} = ue^{-u},$$

for $u > 0$ and $0 < v < 1$.

(a) $f_U(u) = \int_0^1 ue^{-u}\, dv = ue^{-u}$ for $u > 0$, which is a gamma distribution with parameters 2 and 1.

(b) $f_V(v) = \int_0^\infty ue^{-u}\, du = 1$ for $0 < v < 1$. This is a uniform $(0,1)$ distribution.

Chapter 8

Fundamental Sampling Distributions and Data Descriptions

8.1 (a) Responses of all people in Richmond who have telephones.

   (b) Outcomes for a large or infinite number of tosses of a coin.

   (c) Length of life of such tennis shoes when worn on the professional tour.

   (d) All possible time intervals for this lawyer to drive from her home to her office.

8.3 (a) $\bar{x} = 2.4$.

   (b) $\bar{x} = 2$.

   (c) $m = 3$.

8.5 (a) $\bar{x} = 3.2$ seconds.

   (b) $\bar{x} = 3.1$ seconds.

8.7 (a) $\bar{x} = 53.75$.

   (b) Modes are 75 and 100.

8.9 (a) Range $= 15 - 5 = 10$.

   (b) $s^2 = \dfrac{n \sum\limits_{i=1}^{n} x_i^2 - (\sum\limits_{i=1}^{n} x_i)^2}{n(n-1)} = \dfrac{(10)(838) - 86^2}{(10)(9)} = 10.933$. Taking the square root, we have $s = 3.307$.

8.11 (a) $s^2 = \dfrac{1}{n-1} \sum\limits_{x=1}^{n} (x_i - \bar{x})^2 = \dfrac{1}{14}[(2 - 2.4)^2 + (1 - 2.4)^2 + \cdots + (2 - 2.4)^2] = 2.971$.

   (b) $s^2 = \dfrac{n \sum\limits_{i=1}^{n} x_i^2 - (\sum\limits_{i=1}^{n} x_i)^2}{n(n-1)} = \dfrac{(15)(128) - 36^2}{(15)(14)} = 2.971$.

8.13 $s^2 = \dfrac{n \sum\limits_{i=1}^{n} x_i^2 - (\sum\limits_{i=1}^{n} x_i)^2}{n(n-1)} = \dfrac{(2)(148.55) - 53.3^2}{(20)(19)} = 0.342$ and hence $s = 0.585$.

8.15 $s^2 = \dfrac{n\sum\limits_{i=1}^{n} x_i^2 - (\sum\limits_{i=1}^{n} x_i)^2}{n(n-1)} = \dfrac{(6)(207)-33^2}{(6)(5)} = 5.1.$

   (a) Multiplying each observation by 3 gives $s^2 = (9)(5.1) = 45.9.$

   (b) Adding 5 to each observation does not change the variance. Hence $s^2 = 5.1.$

8.17 $z_1 = -1.9$, $z_2 = -0.4$. Hence,

$$P(\mu_{\bar{X}} - 1.9\sigma_{\bar{X}} < \bar{X} < \mu_{\bar{X}} - 0.4\sigma_{\bar{X}}) = P(-1.9 < Z < -0.4) = 0.3446 - 0.0287 = 0.3159.$$

8.19 (a) For $n = 64$, $\sigma_{\bar{X}} = 5.6/8 = 0.7$, whereas for $n = 196$, $\sigma_{\bar{X}} = 5.6/14 = 0.4$. Therefore, the variance of the sample mean is reduced from 0.49 to 0.16 when the sample size is increased from 64 to 196.

   (b) For $n = 784$, $\sigma_{\bar{X}} = 5.6/28 = 0.2$, whereas for $n = 49$, $\sigma_{\bar{X}} = 5.6/7 = 0.8$. Therefore, the variance of the sample mean is increased from 0.04 to 0.64 when the sample size is decreased from 784 to 49.

8.21 $\mu_{\bar{X}} = \mu = 240$, $\sigma_{\bar{X}} = 15/\sqrt{40} = 2.372$. Therefore, $\mu_{\bar{X}} \pm 2\sigma_{\bar{X}} = 240 \pm (2)(2.372)$ or from 235.257 to 244.743, which indicates that a value of $x = 236$ milliliters is reasonable and hence the machine needs not be adjusted.

8.23 (a) $\mu = \sum x f(x) = (4)(0.2) + (5)(0.4) + (6)(0.3) + (7)(0.1) = 5.3$, and
$\sigma^2 = \sum (x - \mu)^2 f(x) = (4 - 5.3)^2(0.2) + (5 - 5.3)^2(0.4) + (6 - 5.3)^2(0.3) + (7 - 5.3)^2(0.1) = 0.81.$

   (b) With $n = 36$, $\mu_{\bar{X}} = \mu = 5.3$ and $\sigma_{\bar{X}} = \sigma^2/n = 0.81/36 = 0.0225.$

   (c) $n = 36$, $\mu_{\bar{X}} = 5.3$, $\sigma_{\bar{X}} = 0.9/6 = 0.15$, and $z = (5.5 - 5.3)/0.15 = 1.33$. So,

$$P(\bar{X} < 5.5) = P(Z < 1.33) = 0.9082.$$

8.25 (a) $P(6.4 < \bar{X} < 7.2) = P(-1.8 < Z < 0.6) = 0.6898.$

   (b) $z = 1.04$, $\bar{x} = z(\sigma/\sqrt{n}) + \mu = (1.04)(1/3) + 7 = 7.35.$

8.27 $n = 50$, $\bar{x} = 0.23$ and $\sigma = 0.1$. Now, $z = (0.23 - 0.2)/(0.1/\sqrt{50}) = 2.12$; so

$$P(\bar{X} \geq 0.23) = P(Z \geq 2.12) = 0.0170.$$

Hence the probability of having such observations, given the mean $\mu = 0.20$, is small. Therefore, the mean amount to be 0.20 is not likely to be true.

8.29 $\mu_{\bar{X}_1 - \bar{X}_2} = 72 - 28 = 44$, $\sigma_{\bar{X}_1 - \bar{X}_2} = \sqrt{100/64 + 25/100} = 1.346$ and $z = (44.2 - 44)/1.346 = 0.15$. So, $P(\bar{X}_1 - \bar{X}_2 < 44.2) = P(Z < 0.15) = 0.5596.$

8.31 The normal quantile-quantile plot is shown as

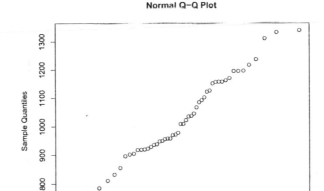

Normal Q-Q Plot

8.33 (a) $n_1 - n_2 = 36$ and $z - 0.2/\sqrt{1/36 + 1/36} = 0.85$. So,

$$P(\bar{X}_B - \bar{X}_A \geq 0.2) = P(Z > 0.85) = 0.1977.$$

(b) Since the probability in (a) is not negligible, the conjecture is not true.

8.35 (a) When the population equals the limit, the probability of a sample mean exceeding the limit would be 1/2 due the symmetry of the approximated normal distribution.

(b) $P(\bar{X} \geq 7960 \mid \mu = 7950) = P(Z \geq (7960 - 7950)/(100/\sqrt{25})) = P(Z \geq 0.5) = 0.3085$. No, this is not very strong evidence that the population mean of the process exceeds the government limit.

8.37 Since the probability that $\bar{X} \leq 775$ is 0.0062, given that $\mu = 800$ is true, it suggests that this event is very rare and it is very likely that the claim of $\mu = 800$ is not true. On the other hand, if μ is truly, say, 760, the probability

$$P(\bar{X} \leq 775 \mid \mu = 760) = P(Z \leq (775 - 760)/(40/\sqrt{16})) = P(Z \leq 1.5) = 0.9332,$$

which is very high.

8.39 (a) 27.488.

(b) 18.475.

(c) 36.415.

8.41 (a) $\chi_\alpha^2 = \chi_{0.99}^2 = 0.297$.

(b) $\chi_\alpha^2 = \chi_{0.025}^2 = 32.852$.

(c) $\chi_{0.05}^2 = 37.652$. Therefore, $\alpha = 0.05 - 0.045 = 0.005$. Hence, $\chi_\alpha^2 = \chi_{0.005}^2 = 46.928$.

8.43 (a) $P(S^2 > 9.1) = P\left(\frac{(n-1)S^2}{\sigma^2} > \frac{(24)(9.1)}{6}\right) = P(\chi^2 > 36.4) = 0.05$.

(b) $P(3.462 < S^2 < 10.745) = P\left(\frac{(24)(3.462)}{6} < \frac{(n-1)S^2}{\sigma^2} < \frac{(24)(10.745)}{6}\right)$

$= P(13.848 < \chi^2 < 42.980) = 0.95 - 0.01 = 0.94.$

8.45 Since $\frac{(n-1)S^2}{\sigma^2}$ is a chi-square statistic, it follows that

$$\sigma^2_{(n-1)S^2/\sigma^2} = \frac{(n-1)^2}{\sigma^4}\sigma^2_{S^2} = 2(n-1).$$

Hence, $\sigma^2_{S^2} = \frac{2\sigma^4}{n-1}$, which decreases as n increases.

8.47 (a) $P(T < 2.365) = 1 - 0.025 = 0.975.$

(b) $P(T > 1.318) = 0.10.$

(c) $P(T < 2.179) = 1 - 0.025 = 0.975$, $P(T < -1.356) = P(T > 1.356) = 0.10$. Therefore, $P(-1.356 < T < 2.179) = 0.975 - 0.010 = 0.875.$

(d) $P(T > -2.567) = 1 - P(T > 2.567) = 1 - 0.01 = 0.99.$

8.49 (a) From Table A.4 we note that 2.069 corresponds to $t_{0.025}$ when $v = 23$. Therefore, $-t_{0.025} = -2.069$ which means that the total area under the curve to the left of $t = k$ is $0.025 + 0.965 = 0.990$. Hence, $k = t_{0.01} = 2.500$.

(b) From Table A.4 we note that 2.807 corresponds to $t_{0.005}$ when $v = 23$. Therefore the total area under the curve to the right of $t = k$ is $0.095 + 0.005 = 0.10$. Hence, $k = t_{0.10} = 1.319$.

8.51 $t = (24 - 20)/(4.1/3) = 2.927$, $t_{0.01} = 2.896$ with 8 degrees of freedom. Conclusion: no, $\mu > 20$.

8.53 (a) 2.71.

(b) 3.51.

(c) 2.92.

(d) $1/2.11 = 0.47.$

(e) $1/2.90 = 0.34.$

8.55 $s_1^2 = 15750$ and $s_2^2 = 10920$ which gives $f = 1.44$. Since, from Table A.6, $f_{0.05}(4,5) = 5.19$, the probability of $F > 1.44$ is much bigger than 0.05, which means that the two variances may be considered equal. The actual probability of $F > 1.44$ is 0.3436 and $P(F < 1/1.44) + P(F > 1.44) = 0.7158$.

Chapter 9

One- and Two-Sample Estimation Problems

9.1 From Example 9.1 on page 271, we know that $E(S^2) = \sigma^2$. Therefore,

$$E(S'^2) = E\left[\frac{n-1}{n}S^2\right] = \frac{n-1}{n}E(S^2) = \frac{n-1}{n}\sigma^2.$$

9.3 $\lim\limits_{n\to\infty} \frac{np+\sqrt{n}/2}{n+\sqrt{n}} = \lim\limits_{n\to\infty} \frac{p+1/2\sqrt{n}}{1+1/\sqrt{n}} = p.$

9.5 $n = 75, \bar{x} = 0.310, \sigma = 0.0015$, and $z_{0.025} = 1.96$. A 95% confidence interval for the population mean is

$$0.310 - (1.96)(0.0015/\sqrt{75}) < \mu < 0.310 + (1.96)(0.0015/\sqrt{75}),$$

or $0.3097 < \mu < 0.3103$.

9.7 $n = 100, \bar{x} = 23,500, \sigma = 3900$, and $z_{0.005} = 2.575$.

   (a) A 99% confidence interval for the population mean is
       $23,500 - (2.575)(3900/10) < \mu < 23,500 + (2.575)(3900/10)$, or
       $22,496 < \mu < 24,504$.

   (b) $e < (2.575)(3900/10) = 1004$.

9.9 $n = [(1.96)(0.0015)/0.0005]^2 = 35$ when rounded up.

9.11 $n = [(2.575)(5.8)/2]^2 = 56$ when rounded up.

9.13 $n = 9, \bar{x} = 1.0056, s = 0.0245$, and $t_{0.005} = 3.355$ with 8 degrees of freedom. A 99% confidence interval for the population mean is

$$1.0056 - (3.355)(0.0245/3) < \mu < 1.0056 + (3.355)(0.0245/3),$$

or $0.978 < \mu < 1.033$.

9.15 $n = 12, \bar{x} = 48.50, s = 1.5$, and $t_{0.05} = 1.796$ with 11 degrees of freedom. A 90% confidence interval for the population mean is

$$48.50 - (1.796)(1.5/\sqrt{12}) < \mu < 48.50 + (1.796)(1.5/\sqrt{12}),$$

or $47.722 < \mu < 49.278$.

9.17 $n = 25, \bar{x} = 325.05, s = 0.5, \gamma = 5\%$, and $1 - \alpha = 90\%$, with $k = 2.208$. So, $325.05 \pm (2.208)(0.5)$ yields $(323.95, 326.151)$. Thus, we are 95% confident that this tolerance interval will contain 90% of the aspirin contents for this brand of buffered aspirin.

9.19 $n = 100, \bar{x} = 23,500, s = 3,900, 1 - \alpha = 0.99$, and $\gamma = 0.01$, with $k = 3.096$. The tolerance interval is $23,500 \pm (3.096)(3,900)$ which yields $11,425 < \mu < 35,574$.

9.21 By definition, $MSE = E(\hat{\Theta} - \theta)^2$ which can be expressed as

$$MSE = E[\hat{\Theta} - E(\hat{\Theta}) + E(\hat{\Theta}) - \theta]^2$$
$$= E[\hat{\Theta} - E(\hat{\Theta})]^2 + E[E(\hat{\Theta}) - \theta]^2 + 2E[\hat{\Theta} - E(\hat{\Theta})]E[E(\hat{\Theta}) - \theta].$$

The third term on the right hand side is zero since $E[\hat{\Theta} - E(\hat{\Theta})] = E[\hat{\Theta}] - E(\hat{\Theta}) = 0$. Hence the claim is valid.

9.23 Using Theorem 8.4, we know that $X^2 = \frac{(n-1)S^2}{\sigma^2}$ follows a chi-squared distribution with $n - 1$ degrees of freedom, whose variance is $2(n-1)$. So $Var(S^2) = Var\left(\frac{\sigma^2}{n-1}X^2\right) = \frac{2}{n-1}\sigma^4$, and $Var(S'^2) = Var\left(\frac{n-1}{n}S^2\right) = \left(\frac{n-1}{n}\right)^2 Var(S^2) = \frac{2(n-1)\sigma^4}{n^2}$. Therefore, the variance of S'^2 is smaller.

9.25 $n = 20, \bar{x} = 11.3, s = 2.45$, and $t_{0.025} = 2.093$ with 19 degrees of freedom. A 95% prediction interval for a future observation is

$$11.3 \pm (2.093)(2.45)\sqrt{1 + 1/20} = 11.3 \pm 5.25,$$

which yields $(6.05, 16.55)$.

9.27 $n = 15, \bar{x} = 3.7867, s = 0.9709$, and $t_{0.025} = 2.145$ with 14 degrees of freedom. A 95% prediction interval for a new observation is

$$3.7867 \pm (2.145)(0.9709)\sqrt{1 + 1/15} = 3.7867 \pm 2.1509,$$

which yields $(1.6358, 5.9376)$.

9.29 $n = 15, \bar{x} = 3.84$, and $s = 3.07$. To calculate an upper 95% prediction limit, we obtain $t_{0.05} = 1.761$ with 14 degrees of freedom. So, the upper limit is $3.84 + (1.761)(3.07)\sqrt{1 + 1/15} = 3.84 + 5.58 = 9.42$. This means that a new observation will have a chance of 95% to fall into the interval $(-\infty, 9.42)$. To obtain an upper 95% tolerance limit, using $1 - \alpha = 0.95$ and $\gamma = 0.05$, with $k = 2.566$, we get $3.84 + (2.566)(3.07) = 11.72$. Hence, we are 95% confident that $(-\infty, 11.72)$ will contain 95% of the orthophosphorous measurements in the river.

9.31 Since the manufacturer would be more interested in the mean tensile strength for future products, it is conceivable that prediction interval and tolerance interval may be more interesting than just a confidence interval.

9.33 In Exercise 9.27, a 95% prediction interval for a new observation is calculated as $(1.6358, 5.9377)$. Since 6.9 is in the outside range of the prediction interval, this new observation is likely to be an outlier.

9.35 $n_1 = 25, n_2 = 36, \bar{x}_1 = 80, \bar{x}_2 = 75, \sigma_1 = 5, \sigma_2 = 3$, and $z_{0.03} = 1.88$. So, a 94% confidence interval for $\mu_1 - \mu_2$ is

$$(80 - 75) - (1.88)\sqrt{25/25 + 9/36} < \mu_1 - \mu_2 < (80 - 75) + (1.88)\sqrt{25/25 + 9/36},$$

which yields $2.9 < \mu_1 - \mu_2 < 7.1$.

9.37 $n_1 = 100, n_2 = 200, \bar{x}_1 = 12.2, \bar{x}_2 = 9.1, s_1 = 1.1$, and $s_2 = 0.9$. It is known that $z_{0.01} = 2.327$. So

$$(12.2 - 9.1) \pm 2.327\sqrt{1.1^2/100 + 0.9^2/200} = 3.1 \pm 0.30,$$

or $2.80 < \mu_1 - \mu_2 < 3.40$. The treatment appears to reduce the mean amount of metal removed.

9.39 $n_1 = 12, n_2 = 18, \bar{x}_1 = 84, \bar{x}_2 = 77, s_1 = 4, s_2 = 6$, and $s_p = 5.305$ with $t_{0.005} = 2.763$ with 28 degrees of freedom. So,

$$(84 - 77) \pm (2.763)(5.305)\sqrt{1/12 + 1/18} = 7 \pm 5.46,$$

which yields $1.54 < \mu_1 - \mu_2 < 12.46$.

9.41 $n_1 = 14, n_2 = 16, \bar{x}_1 = 17, \bar{x}_2 = 19, s_1^2 = 1.5, s_2^2 = 1.8$, and $s_p = 1.289$ with $t_{0.005} = 2.763$ with 28 degrees of freedom. So,

$$(19 - 17) \pm (2.763)(1.289)\sqrt{1/16 + 1/14} = 2 \pm 1.30,$$

which yields $0.70 < \mu_1 - \mu_2 < 3.30$.

9.43 $n_A = n_B = 12, \bar{x}_A = 36,300, \bar{x}_B = 38,100, s_A = 5,000, s_B = 6,100$, and

$$v = \frac{5000^2/12 + 6100^2/12}{\frac{(5000^2/12)^2}{11} + \frac{(6100^2/12)^2}{11}} = 21,$$

with $t_{0.025} = 2.080$ with 21 degrees of freedom. So,

$$(36,300 - 38,100) \pm (2.080)\sqrt{\frac{5000^2}{12} + \frac{6100^2}{12}} = -1,800 \pm 4,736,$$

which yields $-6,536 < \mu_A - \mu_B < 2,936$.

9.45 $n = 9, \bar{d} = 2.778, s_d = 4.5765$, with $t_{0.025} = 2.306$ with 8 degrees of freedom. So,

$$2.778 \pm (2.306)\frac{4.5765}{\sqrt{9}} = 2.778 \pm 3.518,$$

which yields $-0.74 < \mu_D < 6.30$.

9.47 $n = 10, \bar{d} = 14.89\%$, and $s_d = 30.4868$, with $t_{0.025} = 2.262$ with 9 degrees of freedom. So,

$$14.89 \pm (2.262)\frac{30.4868}{\sqrt{10}} = 14.89 \pm 21.81,$$

which yields $-6.92 < \mu_D < 36.70$.

9.49 $n_A = n_B = 15, \bar{x}_A = 3.82, \bar{x}_B = 4.94, s_A = 0.7794, s_B = 0.7538$, and $s_p = 0.7667$ with $t_{0.025} = 2.048$ with 28 degrees of freedom. So,

$$(4.94 - 3.82) \pm (2.048)(0.7667)\sqrt{1/15 + 1/15} = 1.12 \pm 0.57,$$

which yields $0.55 < \mu_B - \mu_A < 1.69$.

9.51 (a) $n = 200, \hat{p} = 0.57, \hat{q} = 0.43$, and $z_{0.02} = 2.05$. So,

$$0.57 \pm (2.05)\sqrt{\frac{(0.57)(0.43)}{200}} = 0.57 \pm 0.072,$$

which yields $0.498 < p < 0.642$.

(b) Error $\leq (2.05)\sqrt{\frac{(0.57)(0.43)}{200}} = 0.072$.

9.53 $n = 1000, \hat{p} = \frac{228}{1000} = 0.228, \hat{q} = 0.772$, and $z_{0.005} = 2.575$. So,

$$0.228 \pm (2.575)\sqrt{\frac{(0.228)(0.772)}{1000}} = 0.228 \pm 0.034,$$

which yields $0.194 < p < 0.262$.

9.55 (a) $n = 40, \hat{p} = \frac{34}{40} = 0.85, \hat{q} = 0.15$, and $z_{0.025} = 1.96$. So,

$$0.85 \pm (1.96)\sqrt{\frac{(0.85)(0.15)}{40}} = 0.85 \pm 0.111,$$

which yields $0.739 < p < 0.961$.

(b) Since $p = 0.8$ falls in the confidence interval, we can not conclude that the new system is better.

9.57 $n = 1600, \hat{p} = \frac{2}{3}, \hat{q} = \frac{1}{3}$, and $z_{0.025} = 1.96$.

(a) $\frac{2}{3} \pm (1.96)\sqrt{\frac{(2/3)(1/3)}{1600}} = \frac{2}{3} \pm 0.023$, which yields $0.644 < p < 0.690$.

(b) Error $\leq (1.96)\sqrt{\frac{(2/3)(1/3)}{1600}} - 0.023$.

9.59 $n = \frac{(2.05)^2(0.57)(0.43)}{(0.02)^2} = 2576$ when round up.

9.61 $n = \frac{(2.33)^2(0.08)(0.92)}{(0.05)^2} = 160$ when round up.

9.63 $n = \frac{(2.575)^2}{(4)(0.01)^2} = 16577$ when round up.

9.65 $n_M = n_F = 1000, \hat{p}_M = 0.250, \hat{q}_M = 0.750, \hat{p}_F = 0.275, \hat{q}_F = 0.725$, and $z_{0.025} = 1.96$. So

$$(0.275 - 0.250) \pm (1.96)\sqrt{\frac{(0.250)(0.750)}{1000} + \frac{(0.275)(0.725)}{1000}} = 0.025 \pm 0.039,$$

which yields $-0.0136 < p_F - p_M < 0.0636$.

9.67 $n_1 = n_2 = 500, \hat{p}_1 = \frac{120}{500} = 0.24, \hat{p}_2 = \frac{98}{500} = 0.196$, and $z_{0.05} = 1.645$. So,

$$(0.24 - 0.196) \pm (1.645)\sqrt{\frac{(0.24)(0.76)}{500} + \frac{(0.196)(0.804)}{500}} = 0.044 \pm 0.0429,$$

which yields $0.0011 < p_1 - p_2 < 0.0869$. Since 0 is not in this confidence interval, we conclude, at the level of 90% confidence, that inoculation has an effect on the incidence of the disease.

9.69 $n_{\text{now}} = 1000, \hat{p}_{\text{now}} = 0.2740, n_{91} = 760, \hat{p}_{91} = 0.3158$, and $z_{0.025} = 1.96$. So,

$$(0.2740 - 0.3158) \pm (1.96)\sqrt{\frac{(0.2740)(0.7260)}{1000} + \frac{(0.3158)(0.6842)}{760}} = -0.0418 \pm 0.0431,$$

which yields $-0.0849 < p_{\text{now}} - p_{91} < 0.0013$. Hence, at the confidence level of 95%, the significance cannot be shown.

9.71 $s^2 = 0.815$ with $v = 4$ degrees of freedom. Also, $\chi^2_{0.025} = 11.143$ and $\chi^2_{0.975} = 0.484$. So,

$$\frac{(4)(0.815)}{11.143} < \sigma^2 < \frac{(4)(0.815)}{0.484}, \quad \text{which yields} \quad 0.293 < \sigma^2 < 6.736.$$

Since this interval contains 1, the claim that σ^2 seems valid.

9.73 $s^2 = 6.0025$ with $v = 19$ degrees of freedom. Also, $\chi^2_{0.025} = 32.852$ and $\chi^2_{0.975} = 8.907$. Hence,

$$\frac{(19)(6.0025)}{32.852} < \sigma^2 < \frac{(19)(6.0025)}{8.907}, \quad \text{or } 3.472 < \sigma^2 < 12.804,$$

9.75 $s^2 = 225$ with $v = 9$ degrees of freedom. Also, $\chi^2_{0.005} = 23.589$ and $\chi^2_{0.995} = 1.735$. Hence,

$$\frac{(9)(225)}{23.589} < \sigma^2 < \frac{(9)(225)}{1.735}, \quad \text{or } 85.845 < \sigma^2 < 1167.147,$$

which yields $9.27 < \sigma < 34.16$.

9.77 $s_1^2 = 1.00, s_2^2 = 0.64, f_{0.01}(11, 9) = 5.19$, and $f_{0.01}(9, 11) = 4.63$. So,

$$\frac{1.00/0.64}{5.19} < \frac{\sigma_1^2}{\sigma_2^2} < (1.00/0.64)(4.63), \text{ or } 0.301 < \frac{\sigma_1^2}{\sigma_2^2} < 7.234,$$

which yields $0.549 < \frac{\sigma_1}{\sigma_2} < 2.690$.

9.79 $s_I^2 = 76.3, s_{II}^2 = 1035.905, f_{0.05}(4, 6) = 4.53$, and $f_{0.05}(6, 4) = 6.16$. So,

$$\left(\frac{76.3}{1035.905}\right)\left(\frac{1}{4.53}\right) < \frac{\sigma_I^2}{\sigma_{II}^2} < \left(\frac{76.3}{1035.905}\right)(6.16), \text{ or } 0.016 < \frac{\sigma_I^2}{\sigma_{II}^2} < 0.454.$$

Hence, we may assume that $\sigma_I^2 \neq \sigma_{II}^2$.

9.81 The likelihood function is

$$L(x_1, \ldots, x_n) = \prod_{i=1}^{n} f(x_i; p) = \prod_{i=1}^{n} p^{x_i}(1-p)^{1-x_i} = p^{n\bar{x}}(1-p)^{n(1-\bar{x})}.$$

Hence, $\ln L = n[\bar{x} \ln(p) + (1 - \bar{x}) \ln(1 - p)]$. Taking derivative with respect to p and setting the derivative to zero, we obtain $\frac{\partial \ln(L)}{\partial p} = n\left(\frac{\bar{x}}{p} - \frac{1-\bar{x}}{1-p}\right) = 0$, which yields $\frac{\bar{x}}{p} - \frac{1-\bar{x}}{1-p} = 0$. Therefore, $\hat{p} = \bar{x}$.

9.83 (a) The likelihood function is

$$L(x_1, \ldots, x_n) = \prod_{i=1}^{n} f(x_i; \mu, \sigma) = \prod_{i=1}^{n} \left\{ \frac{1}{\sqrt{2\pi} \, \sigma x_i} e^{-\frac{[\ln(x_i) - \mu]^2}{2\sigma^2}} \right\}$$

$$= \frac{1}{(2\pi)^{n/2}\sigma^n \prod_{i=1}^{n} x_i} \exp\left\{ -\frac{1}{2\sigma^2} \sum_{i=1}^{n} [\ln(x_i) - \mu]^2 \right\}.$$

(b) It is easy to obtain

$$\ln L = -\frac{n}{2} \ln(2\pi) - \frac{n}{2} \ln(\sigma^2) - \sum_{i=1}^{n} \ln(x_i) - \frac{1}{2\sigma^2} \sum_{i=1}^{n} [\ln(x_i) - \mu]^2.$$

So, setting $0 = \frac{\partial \ln L}{\partial \mu} = \frac{1}{\sigma^2} \sum_{i=1}^{n} [\ln(x_i) - \mu]$, we obtain $\hat{\mu} = \frac{1}{n} \sum_{i=1}^{n} \ln(x_i)$, and setting $0 = \frac{\partial \ln L}{\partial \sigma^2} = -\frac{n}{2\sigma^2} + \frac{1}{2\sigma^4} \sum_{i=1}^{n} [\ln(x_i) - \mu]^2$, we get $\hat{\sigma}^2 = \frac{1}{n} \sum_{i=1}^{n} [\ln(x_i) - \hat{\mu}]^2$.

9.85 $L(x) = p^x(1-p)^{1-x}$, and $\ln L = x \ln(p) + (1-x) \ln(1-p)$, with $\frac{\partial \ln L}{\partial p} = \frac{x}{p} - \frac{1-x}{1-p} = 0$, we obtain $\hat{p} = x = 1$.

Chapter 10

One- and Two-Sample Tests of Hypotheses

10.1 (a) Conclude that fewer than 30% of the public are allergic to some cheese products when, in fact, 30% or more are allergic.

   (b) Conclude that at least 30% of the public are allergic to some cheese products when, in fact, fewer than 30% are allergic.

10.3 (a) The firm is not guilty.

   (b) The firm is guilty.

10.5 (a) $\alpha = P(X < 110 \mid p = 0.6) + P(X > 130 \mid p = 0.6) = P(Z < -1.52) + P(Z > 1.52) = 2(0.0643) = 0.1286.$

   (b) $\beta = P(110 < X < 130 \mid p = 0.5) = P(1.34 < Z < 4.31) = 0.0901.$
   $\beta = P(110 < X < 130 \mid p = 0.7) = P(-4.71 < Z < -1.47) = 0.0708.$

   (c) The probability of a Type I error is somewhat high for this procedure, although Type II errors are reduced dramatically.

10.7 (a) $\alpha = P(X \le 24 \mid p = 0.6) = P(Z < -1.59) = 0.0559.$

   (b) $\beta = P(X > 24 \mid p = 0.3) = P(Z > 2.93) = 1 - 0.9983 = 0.0017.$
   $\beta = P(X > 24 \mid p = 0.4) = P(Z > 1.30) = 1 - 0.9032 = 0.0968.$
   $\beta = P(X > 24 \mid p = 0.5) = P(Z > -0.14) = 1 - 0.4443 = 0.5557.$

10.9 (a) $n = 100$, $p = 0.7$, $\mu = np = 70$, and $\sigma = \sqrt{npq} = \sqrt{(100)(0.7)(0.3)} = 4.583$. Hence $z = \frac{82.5 - 70}{4.583} = 0.3410$. Therefore,

$$\alpha = P(X > 82) = P(Z > 2.73) = 1 - 0.9968 = 0.0032.$$

   (b) $n = 100$, $p = 0.9$, $\mu = np = 90$, and $\sigma = \sqrt{npq} = \sqrt{(100)(0.9)(0.1)} = 3$. Hence $z = \frac{82.5 - 90}{3} = -2.5$. So,

$$\beta = P(X \le 82) = P(X < -2.5) = 0.0062.$$

10.11 (a) $n = 70$, $p = 0.4$, $\mu = np = 28$, and $\sigma = \sqrt{npq} = 4.099$, with $z = \frac{23.5-28}{4.099} = -1.10$.
Then $\alpha = P(X < 24) = P(Z < -1.10) = 0.1357$.

(b) $n = 70$, $p = 0.3$, $\mu = np = 21$, and $\sigma = \sqrt{npq} = 3.834$, with $z = \frac{23.5-21}{3.834} = 0.65$
Then $\beta = P(X \geq 24) = P(Z > 0.65) = 0.2578$.

10.13 From Exercise 10.12(a) we have $\mu = 240$ and $\sigma = 9.798$. We then obtain

$$z_1 = \frac{214.5 - 240}{9.978} = -2.60, \quad \text{and} \quad z_2 = \frac{265.5 - 240}{9.978} = 2.60.$$

So

$$\alpha = 2P(Z < -2.60) = (2)(0.0047) = 0.0094.$$

Also, from Exercise 10.12(b) we have $\mu = 192$ and $\sigma = 9.992$, with

$$z_1 = \frac{214.5 - 192}{9.992} = 2.25, \quad \text{and} \quad z_2 = \frac{265.5 - 192}{9.992} = 7.36.$$

Therefore,

$$\beta = P(2.25 < Z < 7.36) = 1 - 0.9878 = 0.0122.$$

10.15 (a) $\mu = 200$, $n = 9$, $\sigma = 15$ and $\sigma_{\bar{X}} = \frac{15}{3} = 5$. So,

$$z_1 = \frac{191 - 200}{5} = -1.8, \quad \text{and} \quad z_2 = \frac{209 - 200}{5} = 1.8,$$

with $\alpha = 2P(Z < -1.8) = (2)(0.0359) = 0.0718$.

(b) If $\mu = 215$, then $z - 1 = \frac{191-215}{5} = -4.8$ and $z_2 = \frac{209-215}{5} = -1.2$, with

$$\beta = P(-4.8 < Z < -1.2) = 0.1151 - 0 = 0.1151.$$

10.17 (a) $n = 50$, $\mu = 5000$, $\sigma = 120$, and $\sigma_{\bar{X}} = \frac{120}{\sqrt{50}} = 16.971$, with $z = \frac{4970-5000}{16.971} = -1.77$
and $\alpha = P(Z < -1.77) = 0.0384$.

(b) If $\mu = 4970$, then $z = 0$ and hence $\beta = P(Z > 0) = 0.5$.
If $\mu = 4960$, then $z = \frac{4970-4960}{16.971} = 0.59$ and $\beta = P(Z > 0.59) = 0.2776$.

10.19 The hypotheses are

$$H_0 : \mu = 800,$$
$$H_1 : \mu \neq 800.$$

Now, $z = \frac{788-800}{40/\sqrt{30}} = -1.64$, and P-value$= 2P(Z < -1.64) = (2)(0.0505) = 0.1010$.
Hence, the mean is not significantly different from 800 for $\alpha < 0.101$.

10.21 The hypotheses are

$$H_0 : \mu = 40 \text{ months},$$
$$H_1 : \mu < 40 \text{ months}.$$

Now, $z = \frac{38-40}{5.8/\sqrt{64}} = -2.76$, and P-value$= P(Z < -2.76) = 0.0029$. Decision: reject H_0.

10.23 The hypotheses are

$$H_0 : \mu = 20,000 \text{ kilometers},$$
$$H_1 : \mu > 20,000 \text{ kilometers}.$$

Now, $z = \frac{23,500-20,000}{3900/\sqrt{100}} = 8.97$, and P-value$= P(Z > 8.97) \approx 0$. Decision: reject H_0 and conclude that $\mu \neq 20,000$ kilometers.

10.25 The hypotheses are

$$H_0 : \mu = 10,$$
$$H_1 : \mu \neq 10.$$

$\alpha = 0.01$ and $df = 9$.
Critical region: $t < -3.25$ or $t > 3.25$.
Computation: $t = \frac{10.06-10}{0.246/\sqrt{10}} = 0.77$.
Decision: Fail to reject H_0.

10.27 The hypotheses are

$$H_0 : \mu_1 = \mu_2,$$
$$H_1 : \mu_1 > \mu_2.$$

Since $s_p = \sqrt{\frac{(29)(10.5)^2+(29)(10.2)^2}{58}} = 10.35$, then

$$P\left[T > \frac{34.0}{10.35\sqrt{1/30 + 1/30}}\right] = P(Z > 12.72) \approx 0.$$

Hence, the conclusion is that running increases the mean RMR in older women.

10.29 The hypotheses are

$$H_0 : \mu = 35 \text{ minutes},$$
$$H_1 : \mu < 35 \text{ minutes}.$$

$\alpha = 0.05$ and $df = 19$.
Critical region: $t < -1.729$.
Computation: $t = \frac{33.1-35}{4.3/\sqrt{20}} = -1.98$.
Decision: Reject H_0 and conclude that it takes less than 35 minutes, on the average, to take the test.

10.31 The hypotheses are

$$H_0 : \mu_A - \mu_B = 12 \text{ kilograms},$$
$$H_1 : \mu_A - \mu_B > 12 \text{ kilograms}.$$

$\alpha = 0.05$.

Critical region: $z > 1.645$.

Computation: $z = \frac{(86.7-77.8)-12}{\sqrt{(6.28)^2/50+(5.61)^2/50}} = -2.60$. So, fail to reject H_0 and conclude that the average tensile strength of thread A does not exceed the average tensile strength of thread B by 12 kilograms.

10.33 The hypotheses are

$$H_0 : \mu_1 - \mu_2 = 0.5 \text{ micromoles per 30 minutes},$$
$$H_1 : \mu_1 - \mu_2 > 0.5 \text{ micromoles per 30 minutes}.$$

$\alpha = 0.01$.

Critical region: $t > 2.485$ with 25 degrees of freedom.

Computation: $s_p^2 = \frac{(14)(1.5)^2+(11)(1.2)^2}{25} = 1.8936$, and $t = \frac{(8.8-7.5)-0.5}{\sqrt{1.8936}\sqrt{1/15+1/12}} = 1.50$. Do not reject H_0.

10.35 The hypotheses are

$$H_0 : \mu_1 - \mu_2 = 0,$$
$$H_1 : \mu_1 - \mu_2 < 0.$$

$\alpha = 0.05$

Critical region: $t < -1.895$ with 7 degrees of freedom.

Computation: $s_p = \sqrt{\frac{(3)(1.363)+(4)(3.883)}{7}} = 1.674$, and $t = \frac{2.075-2.860}{1.674\sqrt{1/4+1/5}} = -0.70$.

Decision: Do not reject H_0.

10.37 The hypotheses are

$$H_0 : \mu_1 - \mu_2 = 4 \text{ kilometers},$$
$$H_1 : \mu_1 - \mu_2 \neq 4 \text{ kilometers}.$$

$\alpha = 0.10$ and the critical regions are $t < -1.725$ or $t > 1.725$ with 20 degrees of freedom.

Computation: $t = \frac{5-4}{(0.915)\sqrt{1/12+1/10}} = 2.55$.

Decision: Reject H_0.

10.39 The hypotheses are

$$H_0 : \mu_{II} - \mu_I = 10,$$
$$H_1 : \mu_{II} - \mu_I > 10.$$

$\alpha = 0.1$.

Degrees of freedom is calculated as

$$v = \frac{(78.8/5 + 913.333/7)^2}{(78.8/5)^2/4 + (913/333/7)^2/6} = 7.38,$$

hence we use 7 degrees of freedom with the critical region $t > 2.998$.

Computation: $t = \frac{(110-97.4)-10}{\sqrt{78.800/5+913.333/7}} = 0.22$.

Decision: Fail to reject H_0.

10.41 The hypotheses are

$$H_0 : \mu_1 = \mu_2,$$
$$H_1 : \mu_1 \neq \mu_2.$$

$\alpha = 0.05$.

Degrees of freedom is calculated as

$$v = \frac{(7874.329^2/16 + 2479/503^2/12)^2}{(7874.329^2/16)^2/15 + (2479.503^2/12)^2/11} = 19 \text{ degrees of freedom.}$$

Critical regions $t < -2.093$ or $t > 2.093$.

Computation: $t = \frac{9897.500-4120.833}{\sqrt{7874.329^2/16+2479.503^2/12}} = 2.76$.

Decision: Reject H_0 and conclude that $\mu_1 > \mu_2$.

10.43 The hypotheses are

$$H_0 : \mu_1 = \mu_2,$$
$$H_1 : \mu_1 > \mu_2.$$

Computation: $\bar{d} = 0.1417$, $s_d = 0.198$, $t = \frac{0.1417}{0.198/\sqrt{12}} = 2.48$ and $0.015 < P\text{-value} < 0.02$ with 11 degrees of freedom.

Decision: Reject H_0 when a significance level is above 0.02.

10.45 The hypotheses are

$$H_0 : \mu_1 = \mu_2,$$
$$H_1 : \mu_1 < \mu_2.$$

Computation: $\bar{d} = -54.13$, $s_d = 83.002$, $t = \frac{-54.13}{83.002/\sqrt{15}} = -2.53$, and $0.01 < P\text{-value} < 0.015$ with 14 degrees of freedom.

Decision: Reject H_0.

10.47 $n = \frac{(1.645+1.282)^2(0.24)^2}{0.3^2} = 5.48$. The sample size needed is 6.

10.49 $1 - \beta = 0.95$ so $\beta = 0.05$, $\delta = 3.1$ and $z_{0.01} = 2.33$. Therefore,

$$n = \frac{(1.645 + 2.33)^2(6.9)^2}{3.1^2} = 78.28 \approx 79 \text{ due to round up.}$$

10.51 $n = \frac{1.645+0.842)^2(2.25)^2}{[(1.2)(2.25)]^2} = 4.29$. The sample size would be 5.

10.53 (a) The hypotheses are

$$H_0 : M_{\text{hot}} - M_{\text{cold}} = 0.$$
$$H_1 : M_{\text{hot}} - M_{\text{cold}} \neq 0.$$

(b) Use paired T-test and find out $t = 0.99$ with $0.3 < P$-value < 0.4. Hence, fail to reject H_0.

10.55 The hypotheses are

$$H_0 : p = 0.40,$$
$$H_1 : p > 0.40.$$

Denote by X for those who choose lasagna.

$$P\text{-value} = P(X \geq 9 \mid p = 0.40) = 0.4044.$$

The claim that $p = 0.40$ is not refuted.

10.57 The hypotheses are

$$H_0 : p = 0.5,$$
$$H_1 : p < 0.5.$$

$$P\text{-value} = P(X \leq 5 \mid p = 0.05) = 0.0207.$$

Decision: Reject H_0.

10.59 The hypotheses are

$$H_0 : p = 0.2,$$
$$H_1 : p < 0.2.$$

Then

$$P\text{-value} \approx P\left(Z < \frac{136 - (1000)(0.2)}{\sqrt{(1000)(0.2)(0.8)}} \right) = F(Z < -5.06) \approx 0.$$

Decision: Reject H_0; less than $1/5$ of the homes in the city are heated by oil.

10.61 The hypotheses are

$$H_0 : p = 0.8,$$
$$H_1 : p > 0.8.$$

$\alpha = 0.04$.
Critical region: $z > 1.75$.
Computation: $z = \frac{250 - (300)(0.8)}{\sqrt{(300)(0.8)(0.2)}} = 1.44$.
Decision: Fail to reject H_0; it cannot conclude that the new missile system is more accurate.

10.63 The hypotheses are

$$H_0 : p_1 = p_2,$$
$$H_1 : p_1 \neq p_2.$$

Computation: $\hat{p} = \frac{63+59}{100+125} = 0.5422$, $z = \frac{(63/100)-(59/125)}{\sqrt{(0.5422)(0.4578)(1/100+1/125)}} = 2.36$, with
P-value $= 2P(Z > 2.36) = 0.0182$.
Decision: Reject H_0 at level 0.0182. The proportion of urban residents who favor the nuclear plant is larger than the proportion of suburban residents who favor the nuclear plant.

10.65 The hypotheses are

$$H_0 : p_U = p_R,$$
$$H_1 : p_U > p_R.$$

Computation: $\hat{p} = \frac{20+10}{200+150} = 0.085714$, $z = \frac{(20/200)-(10/150)}{\sqrt{(0.085714)(0.914286)(1/200+1/150)}} = 1.10$, with
P-value $= P(Z > 1.10) = 0.1357$.
Decision: Fail to reject H_0. It cannot be shown that breast cancer is more prevalent in the urban community.

10.67 The hypotheses are

$$H_0 : \sigma^2 = 0.03,$$
$$H_1 : \sigma^2 \neq 0.03.$$

Computation: $\chi^2 = \frac{(9)(0.24585)^2}{0.03} = 18.13$. Since $0.025 < P(\chi^2 > 18.13) < 0.05$ with 9 degrees of freedom, $0.05 < P$-value $= 2P(\chi^2 > 18.13) < 0.10$.
Decision: Fail to reject H_0; the sample of 10 containers is not sufficient to show that σ^2 is not equal to 0.03.

10.69 The hypotheses are

$$H_0 : \sigma^2 = 4.2 \text{ ppm},$$
$$H_1 : \sigma^2 \neq 4.2 \text{ ppm}.$$

Computation: $\chi^2 = \frac{(63)(4.25)^2}{4.2} = 63.75$. Since $0.3 < P(\chi^2 > 63.75) < 0.5$ with 63 degrees of freedom, P-value $= 2P(\chi^2 > 18.13) > 0.6$ (In Microsoft Excel, if you type "=2*chidist(63.75,63)", you will get the P-value as 0.8898.
Decision: Fail to reject H_0; the variance of aflotoxins is not significantly different from 4.2 ppm.

10.71 The hypotheses are

$$H_0 : \sigma^2 = 1.15,$$
$$H_1 : \sigma^2 > 1.15.$$

Computation: $\chi^2 = \frac{(24)(2.03)^2}{1.15} = 42.37$. Since $0.01 < P(\chi^2 > 42.37) < 0.02$ with 24 degrees of freedom, $0.01 < P\text{-value} < 0.02$.

Decision: Reject H_0; there is sufficient evidence to conclude, at level $\alpha = 0.05$, that the soft drink machine is out of control.

10.73 The hypotheses are

$$H_0 : \sigma_1^2 = \sigma_2^2,$$
$$H_1 : \sigma_1^2 > \sigma_2^2.$$

Computation: $f = \frac{(6.1)^2}{(5.3)^2} = 1.33$. Since $f_{0.05}(10, 13) = 2.67 > 1.33$, we fail to reject H_0 at level $\alpha = 0.05$. So, the variability of the time to assemble the product is not significantly greater for men. On the other hand, if you use "=fdist(1.33,10,13)", you will obtain the P-value $= 0.3095$.

10.75 The hypotheses are

$$H_0 : \sigma_1^2 = \sigma_2^2,$$
$$H_1 : \sigma_1^2 \neq \sigma_2^2.$$

Computation: $f = \frac{78.800}{913.333} = 0.086$. Since $P\text{-value} = 2P(f < 0.086) = (2)(0.0164) = 0.0328$ for 4 and 6 degrees of freedom, the variability of running time for company 1 is significantly less than, at level 0.0328, the variability of running time for company 2.

10.77 The hypotheses are

$$H_0 : \sigma_1 = \sigma_2,$$
$$H_1 : \sigma_1 \neq \sigma_2.$$

Computation: $f = \frac{(0.0553)^2}{(0.0125)^2} = 19.67$. Since $P\text{-value} = 2P(f > 19.67) = (2)(0.0004) = 0.0008$ for 7 and 7 degrees of freedom, production line 1 is not producing as consistently as production 2.

10.79 The hypotheses are

$$H_0 : \text{die is balanced},$$
$$H_1 : \text{die is unbalanced}.$$

$\alpha = 0.01$.

Critical region: $\chi^2 > 15.086$ with 5 degrees of freedom.

Computation: Since $e_i = 30$, for $i = 1, 2, \ldots, 6$, then

$$\chi^2 = \frac{(28 - 30)^2}{30} + \frac{(36 - 30)^2}{30} + \cdots + \frac{(23 - 30)^2}{30} = 4.47.$$

Decision: Fail to reject H_0; the die is balanced.

10.81 The hypotheses are

$$H_0 : \text{nuts are mixed in the ratio 5:2:2:1,}$$
$$H_1 : \text{nuts are not mixed in the ratio 5:2:2:1.}$$

$\alpha = 0.05$.
Critical region: $\chi^2 > 7.815$ with 3 degrees of freedom.
Computation:

Observed	269	112	74	45
Expected	250	100	100	50

$$\chi^2 = \frac{(269 - 250)^2}{250} + \frac{(112 - 100)^2}{100} + \frac{(74 - 100)^2}{100} + \frac{(45 - 50)^2}{50} = 10.14.$$

Decision: Reject H_0; the nuts are not mixed in the ratio 5:2:2:1.

10.83 The hypotheses are

$$H_0 : \text{Data follows the binomial distribution } b(y; 3, 1/4),$$
$$H_1 : \text{Data does not follows the binomial distribution.}$$

$\alpha = 0.01$.
Computation: $b(0; 3, 1/4) = 27/64$, $b(1; 3, 1/4) = 27/64$, $b(2; 3, 1/4) = 9/64$, and $b(3; 3, 1/4) = 1/64$. Hence $e_1 = 27$, $e_2 = 27$, $e_3 = 9$ and $e_4 = 1$. Combining the last two classes together, we obtain

$$\chi^2 = \frac{(21 - 27)^2}{27} + \frac{(31 - 27)^2}{27} + \frac{(12 - 10)^2}{10} = 2.33.$$

Critical region: $\chi^2 > 9.210$ with 2 degrees of freedom.
Decision: Fail to reject H_0; the data is from a distribution not significantly different from $b(y; 3, 1/4)$.

10.85 The hypotheses are

$$H_0 : f(x) = g(x; 1/2) \text{ for } x = 1, 2, \ldots,$$
$$H_1 : f(x) \neq g(x; 1/2).$$

$\alpha = 0.05$.
Computation: $g(x; 1/2) = \frac{1}{2^x}$, for $x = 1, 2, \ldots, 7$ and $P(X \geq 8) = \frac{1}{2^7}$. Hence $e_1 = 128$, $e_2 = 64$, $e_3 = 32$, $e_4 = 16$, $e_5 = 8$, $e_6 = 4$, $e_7 = 2$ and $e_8 = 2$. Combining the last three classes together, we obtain

$$\chi^2 = \frac{(136 - 128)^2}{128} + \frac{(60 - 64)^2}{64} + \frac{(34 - 32)^2}{32} + \frac{(12 - 16)^2}{16} + \frac{(9 - 8)^2}{8} + \frac{(5 - 8)^2}{8} = 3.125$$

Critical region: $\chi^2 > 11.070$ with 5 degrees of freedom.
Decision: Fail to reject H_0; $f(x) = g(x; 1/2)$, for $x = 1, 2, \ldots$

10.89 From the data we have

z values	$P(Z < z)$	$P(z_{i-1} < Z < z_i)$	e_i	o_i
$z_1 = \frac{0.795-1.8}{0.4} = -2.51$	0.0060	0.0060	0.2 ⎫	1 ⎫
$z_2 = \frac{0.995-1.8}{0.4} = -2.01$	0.0222	0.0162	0.6 ⎬ 6.1	1 ⎬ 5
$z_3 = \frac{1.195-1.8}{0.4} = -1.51$	0.0655	0.0433	1.7 ⎪	1 ⎪
$z_4 = \frac{1.395-1.8}{0.4} = -1.01$	0.1562	0.0907	3.6 ⎭	2 ⎭
$z_5 = \frac{1.595-1.8}{0.4} = -0.51$	0.3050	0.1488	6.0	4
$z_6 = \frac{1.795-1.8}{0.4} = -0.01$	0.4960	0.1910	7.6	13
$z_7 = \frac{1.995-1.8}{0.4} = 0.49$	0.6879	0.1919	7.7	8
$z_8 = \frac{2.195-1.8}{0.4} = 0.99$	0.8389	0.1510	6.0	5
$z_9 = \frac{2.395-1.8}{0.4} = 1.49$	0.9319	0.0930	3.7 ⎫ 6.4	3 ⎫ 5
$z_{10} = \infty$	1.0000	0.0681	2.7 ⎭	2 ⎭

The hypotheses are

H_0 : Distribution of nicotine contents is normal $n(x; 1.8, 0.4)$,

H_1 : Distribution of nicotine contents is not normal.

$\alpha = 0.01$.

Computation: A goodness-of-fit test with 5 degrees of freedom is based on the following data:

o_i	5	4	13	8	5	5
e_i	6.1	6.0	7.6	7.7	6.0	6.4

Critical region: $\chi^2 > 15.086$.

$$\chi^2 = \frac{(5-6.1)^2}{6.1} + \frac{(4-6.0)^2}{6.0} + \cdots + \frac{(5-6.4)^2}{6.4} = 5.19.$$

Decision: Fail to reject H_0; distribution of nicotine contents is not significantly different from $n(x; 1.8, 0.4)$.

10.91 The hypotheses are

H_0 : A person's gender and time spent watching television are independent,

H_1 : A person's gender and time spent watching television are not independent.

$\alpha = 0.01$.

Critical region: $\chi^2 > 6.635$ with 1 degrees of freedom.

Computation:

Observed and expected frequencies			
	Male	Female	Total
Over 25 hours	15 (20.5)	29 (23.5)	44
Under 25 hours	27 (21.5)	19 (24.5)	46
Total	42	48	90

$$\chi^2 = \frac{(15-20.5)^2}{20.5} + \frac{(29-23.5)^2}{23.5} + \frac{(27-21.5)^2}{21.5} + \frac{(19-24.5)^2}{24.5} = 5.47.$$

Decision: Fail to reject H_0; a person's gender and time spent watching television are independent.

10.93 The hypotheses are

H_0 : Occurrence of types of crime is independent of city district,

H_1 : Occurrence of types of crime is dependent upon city district.

$\alpha = 0.01$.

Critical region: $\chi^2 > 21.666$ with 9 degrees of freedom.

Computation:

Observed and expected frequencies					
District	Assault	Burglary	Larceny	Homicide	Total
1	162 (186.4)	118 (125.8)	451 (423.5)	18 (13.3)	749
2	310 (380.0)	196 (256.6)	996 (863.4)	25 (27.1)	1527
3	258 (228.7)	193 (154.4)	458 (519.6)	10 (16.3)	919
4	280 (214.9)	175 (145.2)	390 (488.5)	19 (15.3)	864
Total	1010	682	2295	72	4059

$$\chi^2 = \frac{(162-186.4)^2}{186.4} + \frac{(118-125.8)^2}{125.8} + \cdots + \frac{(19-15.3)^2}{15.3} = 124.59.$$

Decision: Reject H_0; occurrence of types of crime is dependent upon city district.

10.95 The hypotheses are

H_0 : The attitudes among the four counties are homogeneous,

H_1 : The attitudes among the four counties are not homogeneous.

Computation:

Observed and expected frequencies					
	County				
Attitude	Craig	Giles	Franklin	Montgomery	Total
Favor	65 (74.5)	66 (55.9)	40 (37.3)	34 (37.3)	205
Oppose	42 (53.5)	30 (40.1)	33 (26.7)	42 (26.7)	147
No Opinion	93 (72.0)	54 (54.0)	27 (36.0)	24 (36.0)	198
Total	200	150	100	100	550

$$\chi^2 = \frac{(65-74.5)^2}{74.5} + \frac{(66-55.9)^2}{55.9} + \cdots + \frac{(24-36.0)^2}{36.0} = 31.17.$$

Since P-value $= P(\chi^2 > 31.17) < 0.001$ with 6 degrees of freedom, we reject H_0 and conclude that the attitudes among the four counties are not homogeneous.

10.97 The hypotheses are

H_0 : Proportions of household within each standard of living category are equal,

H_1 : Proportions of household within each standard of living category are not equal.

$\alpha = 0.05$.

Critical region: $\chi^2 > 12.592$ with 6 degrees of freedom.

Computation:

Observed and expected frequencies				
Period	Somewhat Better	Same	Not as Good	Total
1980: Jan.	72 (66.6)	144 (145.2)	84 (88.2)	300
May.	63 (66.6)	135 (145.2)	102 (88.2)	300
Sept.	47 (44.4)	100 (96.8)	53 (58.8)	200
1981: Jan.	40 (44.4)	105 (96.8)	55 (58.8)	200
Total	222	484	294	1000

$$\chi^2 = \frac{(72 - 66.6)^2}{66.6} + \frac{(144 - 145.2)^2}{145.2} + \cdots + \frac{(55 - 58.8)^2}{58.8} = 5.92.$$

Decision: Fail to reject H_0; proportions of household within each standard of living category are equal.

10.99 The hypotheses are

H_0 : Proportions of voters favoring candidate A, candidate B, or undecided are the same for each city,

H_1 : Proportions of voters favoring candidate A, candidate B, or undecided are not the same for each city.

$\alpha = 0.05$.

Critical region: $\chi^2 > 5.991$ with 2 degrees of freedom.

Computation:

Observed and expected frequencies			
	Richmond	Norfolk	Total
Favor A	204 (214.5)	225 (214.5)	429
Favor B	211 (204.5)	198 (204.5)	409
Undecided	85 (81)	77 (81)	162
Total	500	500	1000

$$\chi^2 = \frac{(204 - 214.5)^2}{214.5} + \frac{(225 - 214.5)^2}{214.5} + \cdots + \frac{(77 - 81)^2}{81} = 1.84.$$

Decision: Fail to reject H_0; the proportions of voters favoring candidate A, candidate B, or undecided are not the same for each city.

Chapter 11

Simple Linear Regression and Correlation

11.1 (a) $\sum_i x_i = 778.7$, $\sum_i y_i = 2050.0$, $\sum_i x_i^2 = 26,591.63$, $\sum_i x_i y_i = 65,164.04$, $n = 25$.
Therefore,

$$b = \frac{(25)(65,164.04) - (778.7)(2050.0)}{(25)(26,591.63) - (778.7)^2} = 0.5609,$$

$$a = \frac{2050 - (0.5609)(778.7)}{25} = 64.53.$$

(b) Using the equation $\hat{y} = 64.53 + 0.5609x$ with $x = 30$, we find $\hat{y} = 64.53 + (0.5609)(30) = 81.40$.

(c) Residuals appear to be random as desired.

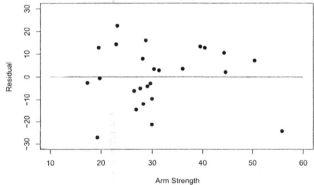

11.3 (a) $\sum_i x_i = 16.5$, $\sum_i y_i = 100.4$, $\sum_i x_i^2 = 25.85$, $\sum_i x_i y_i = 152.59$, $n = 11$. Therefore,

$$b = \frac{(11)(152.59) - (16.5)(100.4)}{(11)(25.85) - (16.5)^2} = 1.8091,$$

$$a = \frac{100.4 - (1.8091)(16.5)}{11} = 6.4136.$$

65

Hence $\hat{y} = 6.4136 + 1.8091x$

(b) For $x = 1.75$, $\hat{y} = 6.4136 + (1.8091)(1.75) = 9.580$.

(c) Residuals appear to be random as desired.

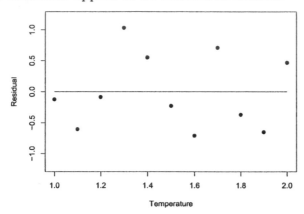

11.5 (a) $\sum_i x_i = 675$, $\sum_i y_i = 488$, $\sum_i x_i^2 = 37,125$, $\sum_i x_i y_i = 25,005$, $n = 18$. Therefore,

$$b = \frac{(18)(25,005) - (675)(488)}{(18)(37,125) - (675)^2} = 0.5676,$$

$$a = \frac{488 - (0.5676)(675)}{18} = 5.8254.$$

Hence $\hat{y} = 5.8254 + 0.5676x$

(b) The scatter plot and the regression line are shown below.

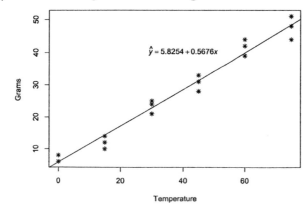

(c) For $x = 50$, $\hat{y} = 5.8254 + (0.5676)(50) = 34.205$ grams.

11.7 (a) The scatter plot and the regression line are shown here.

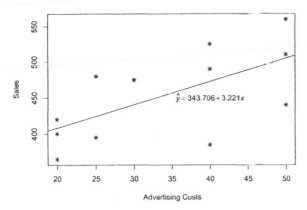

(b) $\sum_i x_i = 410$, $\sum_i y_i = 5445$, $\sum_i x_i^2 = 15,650$, $\sum_i x_i y_i = 191,325$, $n = 12$. Therefore,

$$b = \frac{(12)(191,325) - (410)(5445)}{(12)(15,650) - (410)^2} = 3.2208,$$

$$a = \frac{5445 - (3.2208)(410)}{12} = 343.7056.$$

Hence $\hat{y} = 343.7056 + 3.2208x$

(c) When $x = \$35$, $\hat{y} = 343.7056 + (3.2208)(35) = \$456.43$.

(d) Residuals appear to be random as desired.

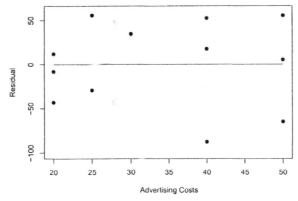

11.9 (a) $\sum_i x_i = 45$, $\sum_i y_i = 1094$, $\sum_i x_i^2 = 244.26$, $\sum_i x_i y_i = 5348.2$, $n = 9$.

$$b = \frac{(9)(5348.2) - (45)(1094)}{(9)(244.26) - (45)^2} = -6.3240,$$

$$a = \frac{1094 - (-6.3240)(45)}{9} = 153.1755.$$

Hence $\hat{y} = 153.1755 - 6.3240x$.

(b) For $x = 4.8$, $\hat{y} = 153.1755 - (6.3240)(4.8) = 123$.

11.11   (a) The scatter plot and the regression line are shown here.

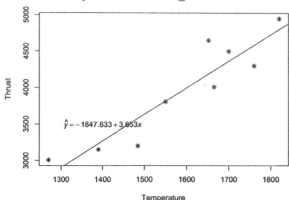

(b) $\sum_i x_i = 14,292$, $\sum_i y_i = 35,578$, $\sum_i x_i^2 = 22,954,054$ $\sum_i x_i y_i = 57,441,610$, $n = 9$. Therefore,

$$b = \frac{(9)(57,441,610) - (14,292)(35,578)}{(9)(22,954,054) - (14,292)^2} = 3.6529,$$

$$a = \frac{35,578 - (3.6529)(14,292)}{9} = -1847.69.$$

Hence $\hat{y} = -1847.69 + 3.6529x$.

11.13   (a) The scatter plot and the regression line are shown here. A simple linear model seems suitable for the data.

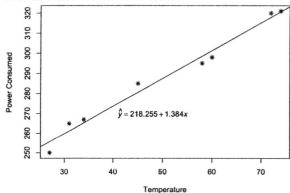

(b) $\sum_i x_i = 999$, $\sum_i y_i = 670$, $\sum_i x_i^2 = 119,969$, $\sum_i x_i y_i = 74,058$, $n = 10$. Therefore,

$$b = \frac{(10)(74,058) - (999)(670)}{(10)(119,969) - (999)^2} = 0.3533,$$

$$a = \frac{670 - (0.3533)(999)}{10} = 31.71.$$

Hence $\hat{y} = 31.71 + 0.3533x$.

(c) See (a).

11.15 The least squares estimator A of α is a linear combination of normally distributed random variables and is thus normal as well.

$$E(A) = E(\bar{Y} - B\bar{x}) = E(\bar{Y}) - \bar{x}E(B) = \alpha + \beta\bar{x} - \beta\bar{x} = \alpha,$$

$$\sigma_A^2 = \sigma_{\bar{Y}-B\bar{x}}^2 = \sigma_{\bar{Y}}^2 + \bar{x}^2\sigma_B^2 - 2\bar{x}\sigma_{\bar{Y}B} = \frac{\sigma^2}{n} + \frac{\bar{x}^2\sigma^2}{n\sum\limits_{i=1}^{n}(x_i - \bar{x})^2}, \text{ since } \sigma_{\bar{Y}B} = 0.$$

Hence

$$\sigma_A^2 = \frac{\sum\limits_{i=1}^{n} x_i^2}{n\sum\limits_{i=1}^{n}(x_i - \bar{x})^2}\sigma^2.$$

11.17 $S_{xx} = 26,591.63 - 778.7^2/25 = 2336.6824$, $S_{yy} = 172,891.46 - 2050^2/25 - 4791.46$, $S_{xy} = 65,164.04 - (778.7)(2050)/25 = 1310.64$, and $b = 0.5609$.

(a) $s^2 = \frac{4791.46-(0.5609)(1310.64)}{23} - 176.362$.

(b) The hypotheses are

$$H_0 : \beta = 0,$$
$$H_1 : \beta \neq 0.$$

$\alpha = 0.05$.

Critical region: $t < -2.069$ or $t > 2.069$.

Computation: $t = \frac{0.5609}{\sqrt{176.362/2336.6824}} = 2.04$.

Decision: Do not reject H_0.

11.19 $S_{xx} = 25.85 - 16.5^2/11 = 1.1$, $S_{yy} = 923.58 - 100.4^2/11 = 7.2018$, $S_{xy} = 152.59 - (165)(100.4)/11 = 1.99$, $a = 6.4136$ and $b = 1.8091$.

(a) $s^2 = \frac{7.2018-(1.8091)(1.99)}{9} = 0.40$.

(b) Since $s = 0.632$ and $t_{0.025} = 2.262$ for 9 degrees of freedom, then a 95% confidence interval is

$$6.4136 \pm (2.262)(0.632)\sqrt{\frac{25.85}{(11)(1.1)}} = 6.4136 \pm 2.0895,$$

which implies $4.324 < \alpha < 8.503$.

(c) $1.8091 \pm (2.262)(0.632)/\sqrt{1.1}$ implies $0.446 < \beta < 3.172$.

11.21 $S_{xx} = 37,125 - 675^2/18 = 11,812.5$, $S_{yy} = 17,142 - 488^2/18 = 3911.7778$, $S_{xy} = 25,005 - (675)(488)/18 = 6705$, $a = 5.8254$ and $b = 0.5676$.

(a) $s^2 = \frac{3911.7778 - (0.5676)(6705)}{16} = 6.626$.

(b) Since $s = 2.574$ and $t_{0.005} = 2.921$ for 16 degrees of freedom, then a 99% confidence interval is

$$5.8261 \pm (2.921)(2.574)\sqrt{\frac{37,125}{(18)(11,812.5)}} = 5.8261 \pm 3.1417,$$

which implies $2.686 < \alpha < 8.968$.

(c) $0.5676 \pm (2.921)(2.574)/\sqrt{11,812.5}$ implies $0.498 < \beta < 0.637$.

11.23 The hypotheses are

$$H_0 : \beta = 6,$$
$$H_1 : \beta < 6.$$

$\alpha = 0.025$.
Critical region: $t = -2.228$.
Computations: $S_{xx} = 15,650 - 410^2/12 = 1641.667$, $S_{yy} = 2,512.925 - 5445^2/12 = 42,256.25$, $S_{xy} = 191,325 - (410)(5445)/12 = 5,287.5$, $s^2 = \frac{42,256.25 - (3,221)(5,287.5)}{10} = 2,522.521$ and then $s = 50.225$. Now

$$t = \frac{3.221 - 6}{50.225/\sqrt{1641.667}} = -2.24.$$

Decision: Reject H_0 and claim $\beta < 6$.

11.25 Using the value $s = 1.64$ from Exercise 11.20(a) and the fact that $y_0 = 25.7724$ when $x_0 = 24.5$, and $\bar{x} = 25.9667$, we have

(a) $25.7724 \pm (2.228)(1.640)\sqrt{\frac{1}{12} + \frac{(-1.4667)^2}{43.0467}} = 25.7724 \pm 1.3341$ implies $24.438 < \mu_{Y \mid 24.5} < 27.106$.

(b) $25.7724 \pm (2.228)(1.640)\sqrt{1 + \frac{1}{12} + \frac{(-1.4667)^2}{43.0467}} = 25.7724 \pm 3.8898$ implies $21.883 < y_0 < 29.662$.

11.27 Using the value $s = 0.632$ from Exercise 11.19(a) and the fact that $y_0 = 9.308$ when $x_0 = 1.6$, and $\bar{x} = 1.5$, we have

$$9.308 \pm (2.262)(0.632)\sqrt{1 + \frac{1}{11} + \frac{0.1^2}{1.1}} = 9.308 \pm 1.4994$$

implies $7.809 < y_0 < 10.807$.

11.29 (a) 17.1812.

(b) The goal of 30 mpg is unlikely based on the confidence interval for mean mpg, $(27.95, 29.60)$.

(c) Based on the prediction interval, the Lexus ES300 should exceed 18 mpg.

11.31 When there are only two data points $x_1 \neq x_2$, using Exercise 11.30 we know that $(y_1 - \hat{y}_1) + (y_2 - \hat{y}_2) = 0$. On the other hand, by the method of least squares on page 395, we also know that $x_1(y_1 - \hat{y}_1) + x_2(y_2 - \hat{y}_2) = 0$. Both of these equations yield $(x_2 - x_1)(y_2 - \hat{y}_2) = 0$ and hence $y_2 - \hat{y}_2 = 0$. Therefore, $y_1 - \hat{y}_1 = 0$. So,

$$SSE = (y_1 - \hat{y}_1)^2 + (y_2 - \hat{y}_2)^2 = 0.$$

Since $R^2 = 1 - \frac{SSE}{SST}$, we have $R^2 = 1$.

11.33 (a) The scatter plot of the data is shown next.

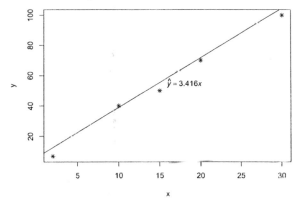

(b) $\sum\limits_{i=1}^{n} x_i^2 = 1629$ and $\sum\limits_{i=1}^{n} x_i y_i = 5564$. Hence $b = \frac{5564}{1629} = 3.4156$. So, $\hat{y} = 3.4156x$.

(c) See (a).

(d) Since there is only one regression coefficient, β, to be estimated, the degrees of freedom in estimating σ^2 is $n - 1$. So,

$$\hat{\sigma}^2 = s^2 = \frac{SSE}{n-1} = \frac{\sum\limits_{i=1}^{n}(y_i - bx_i)^2}{n-1}.$$

(e) $Var(\hat{y}_i) = Var(Bx_i) = x_i^2 Var(B) = \frac{x_i^2 \sigma^2}{\sum\limits_{i=1}^{n} x_i^2}$.

(f) The plot is shown next.

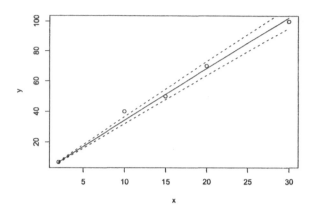

11.35 (a) As shown in Exercise 11.32, the least squares estimator of β is $b = \dfrac{\sum\limits_{i=1}^{n} x_i y_i}{\sum\limits_{i=1}^{n} x_i^2}$.

(b) Since $\sum\limits_{i=1}^{n} x_i y_i = 197.59$, and $\sum\limits_{i=1}^{n} x_i^2 = 98.64$, then $b = \dfrac{197.59}{98.64} = 2.003$ and $\hat{y} = 2.003x$.

11.37 Now since the true model has been changed,

$$E(B) = \frac{\sum\limits_{i=1}^{n} (x_{1i} - \bar{x}_1) E(Y_i)}{\sum\limits_{i=1}^{n} (x_{1i} - \bar{x}_i)^2} = \frac{\sum\limits_{i=1}^{n} (x_{1i} - \bar{x}_1)(\alpha + \beta x_{1i} + \gamma x_{2i})}{\sum\limits_{i=1}^{n} (x_{1i} - \bar{x}_1)^2}$$

$$= \frac{\beta \sum\limits_{i=1}^{n} (x_{1i} - \bar{x}_1)^2 + \gamma \sum\limits_{i=1}^{n} (x_{1i} - \bar{x}_1) x_{2i}}{\sum\limits_{i=1}^{n} (x_{1i} - \bar{x}_1)^2} = \beta + \gamma \frac{\sum\limits_{i=1}^{n} (x_{1i} - \bar{x}_1) x_{2i}}{\sum\limits_{i=1}^{n} (x_{1i} - \bar{x}_1)^2}.$$

11.39 (a) $S_{xx} = 1058$, $S_{yy} = 198.76$, $S_{xy} = -363.63$, $b = \dfrac{S_{xy}}{S_{xx}} = -0.34370$, and $a = \dfrac{210 - (-0.34370)(172.5)}{25} = 10.81153$.

(b) The hypotheses are

$$H_0 : \text{ The regression is linear in } x,$$

$$H_1 : \text{ The regression is nonlinear in } x.$$

$\alpha = 0.05$.

Critical regions: $f > 3.10$ with 3 and 20 degrees of freedom.

Computations: $SST = S_{yy} = 198.76$, $SSR = bS_{xy} = 124.98$ and $SSE = S_{yy} - SSR = 73.98$. Since

$$T_{1.} = 51.1, T_{2.} = 51.5, T_{3.} = 49.3, T_{4.} = 37.0, \text{ and } T_{5.} = 22.1,$$

then

$$SSE(\text{pure}) = \sum_{i=1}^{5} \sum_{j=1}^{5} y_{ij}^2 - \sum_{i=1}^{5} \frac{T_{i.}^2}{5} = 1979.60 - 1910.272 = 69.33.$$

Hence the "Lack-of-fit SS" is $73.78 - 69.33 = 4.45$.

Source of Variation	Sum of Squares	Degrees of Freedom	Mean Square	Computed f
Regression	124.98	1	124.98	
Error	73.98	23	3.22	
$\{$ Lack of fit	$\{$ 4.45	$\{$ 3	$\{$ 1.48	0.43
$\{$ Pure error	$\{$ 69.33	$\{$ 20	$\{$ 3.47	
Total	198.76	24		

Decision: Do not reject H_0.

11.41 The hypotheses are

$$H_0 : \text{The regression is linear in } x,$$

$$H_1 : \text{The regression is nonlinear in } x.$$

$\alpha = 0.05$.

Critical regions: $f > 3.00$ with 6 and 12 degrees of freedom.

Computations: $SST = S_{yy} = 5928.55$, $SSR = bS_{xy} = 1219.35$ and $SSE = S_{yy} - SSR = 4709.20$. $SSE(\text{pure}) = \sum_{i=1}^{8} \sum_{j=1}^{n_i} y_{ij}^2 - \sum_{i=1}^{8} \frac{T_{i.}^2}{n_i} = 3020.67$, and the "Lack-of-fit SS" is $4709.20 - 3020.67 = 1688.53$.

Source of Variation	Sum of Squares	Degrees of Freedom	Mean Square	Computed f
Regression	1219.35	1	1219.35	
Error	4709.20	18	261.62	
$\{$ Lack of fit	$\{$ 1688.53	$\{$ 6	$\{$ 281.42	1.12
$\{$ Pure error	$\{$ 3020.67	$\{$ 12	$\{$ 251.72	
Total	5928.55	19		

Decision: Do not reject H_0; the lack-of-fit test is insignificant.

11.43 $\hat{y} = -21.0280 + 0.4072x$; $f_{\text{LOF}} = 1.71$ with a P-value $= 0.2517$. Hence, lack-of-fit test is insignificant and the linear model is adequate.

11.45 (a) $\hat{y} = -11.3251 - 0.0449$ temperature.

(b) Yes.

(c) 0.9355.

(d) The proportion of impurities does depend on temperature.

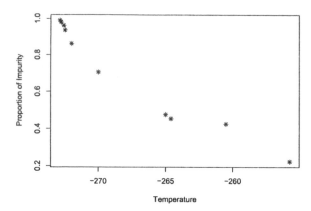

However, based on the plot, it does not appear that the dependence is in linear fashion. If there were replicates, a lack-of-fit test could be performed.

11.47   (a) The figure is shown next.

     (b) $\hat{y} = -175.9025 + 0.0902$ year; $R^2 = 0.3322$.

     (c) There is definitely a relationship between year and nitrogen oxide. It does not appear to be linear.

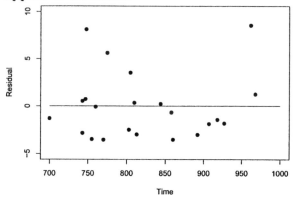

11.49 $S_{xx} = 36,354 - 35,882.667 = 471.333$, $S_{yy} = 38,254 - 37,762.667 = 491.333$, and $S_{xy} = 36,926 - 36,810.667 = 115.333$. So, $r = \frac{115}{\sqrt{(471.333)(491.333)}} = 0.240$.

11.51   Since $b = \frac{S_{xy}}{S_{xx}}$, we can write $s^2 = \frac{S_{yy} - bS_{xy}}{n-2} = \frac{S_{yy} - b^2 S_{xx}}{n-2}$. Also, $b = r\sqrt{\frac{S_{yy}}{S_{xx}}}$ so that $s^2 = \frac{S_{yy} - r^2 S_{yy}}{n-2} = \frac{(1-r^2)S_{yy}}{n-2}$, and hence

$$t = \frac{b}{s\sqrt{S_{xx}}} = \frac{r\sqrt{S_{yy}/S_{xx}}}{\sqrt{S_{yy}S_{xx}(1-r^2)/(n-2)}} = \frac{r\sqrt{n-2}}{\sqrt{1-r^2}}.$$

11.53   (a) From the data of Exercise 11.1 we can calculate

$$S_{xx} = 26,591.63 - (778.7)^2/25 = 2336.6824,$$
$$S_{yy} = 172,891.46 - (2050)^2/25 = 4791.46,$$
$$S_{xy} = 65,164.04 - (778.7)(2050)/25 = 1310.64.$$

Therefore, $r = \dfrac{1310.64}{\sqrt{(2236.6824)(4791.46)}} = 0.392$.

(b) The hypotheses are

$$H_0 : \rho = 0,$$
$$H_1 : \rho \neq 0.$$

$\alpha = 0.05$.

Critical regions: $t < -2.069$ or $t > 2.069$.

Computations: $t = \dfrac{0.392\sqrt{23}}{\sqrt{1-0.392^2}} = 2.04$.

Decision: Fail to reject H_0 at level 0.05. However, the P-value $= 0.0530$ which is marginal.

Chapter 12

Multiple Linear Regression and Certain Nonlinear Regression Models

12.1 (a) $\hat{y} = 27.5467 + 0.9217x_1 + 0.2842x_2$.

   (b) When $x_1 = 60$ and $x_2 = 4$, the predicted value of the chemistry grade is
       $\hat{y} = 27.5467 + (0.9217)(60) + (0.2842)(4) = 84$.

12.3 $\hat{y} = 0.7800 + 2.7122x_1 + 2.0497x_2$.

12.5 (a) $\hat{y} = 56.46333 + 0.15253x - 0.00008x^2$.

   (b) $\hat{y} = 56.46333 + (0.15253)(225) - (0.00008)(225)^2 = 86.73333\%$.

12.7 $\hat{y} = 141.61178 - 0.28193x + 0.00031x^2$.

12.9 (a) $\hat{y} = -102.71324 + 0.60537x_1 + 8.92364x_2 + 1.43746x_3 + 0.01361x_4$.

   (b) $\hat{y} = -102.71324 + (0.60537)(75) + (8.92364)(24) + (1.43746)(90) + (0.01361)(98) = 287.56183$.

12.11 $\hat{y} = 3.3205 + 0.42105x_1 - 0.29578x_2 + 0.01638x_3 + 0.12465x_4$.

12.13 $\hat{y} = -6.51221 + 1.99941x_1 - 3.67510x_2 + 2.52449x_3 + 5.15808x_4 + 14.40116x_5$.

12.15 (a) $\hat{y} = 350.99427 - 1.27199x_1 - 0.15390x_2$.

   (b) $\hat{y} = 350.99427 - (1.27199)(20) - (0.15390)(1200) = 140.86930$.

12.17 $s^2 = 0.16508$.

12.19 $s^2 = 242.71561$.

12.21 Using *SAS* output, we obtain

   (a) $\hat{\sigma}_{b_2}^2 = 28.09554$.

   (b) $\hat{\sigma}_{b_1 b_4} = -0.00958$.

12.23 Using *SAS* output, we obtain a 90% confidence interval for the mean response when $x = 19.5$ as $29.9284 < \mu_{Y|x=19.5} < 31.9729$.

12.25 The hypotheses are

$$H_0 : \beta_2 = 0,$$
$$H_1 : \beta_2 \neq 0.$$

The test statistic value is $t = 2.86$ with a P-value $= 0.0145$. Hence, we reject H_0 and conclude $\beta_2 \neq 0$.

12.27 The hypotheses are

$$H_0 : \beta_1 = 2,$$
$$H_1 : \beta_1 \neq 2.$$

The test statistics is $t = \frac{2.71224 - 2}{0.20209} = 3.524$ with P-value $= 0.0097$. Reject H_0 and conclude that $\beta_1 \neq 2$.

12.29 (a) P-value $= 0.3562$. Hence, fail to reject H_0.

(b) P-value $= 0.1841$. Again, fail to reject H_0.

(c) There is not sufficient evidence that the regressors x_1 and x_2 significantly influence the response with the described linear model.

12.31 $R^2 = \frac{SSR}{SST} = \frac{10953}{10956} = 99.97\%$. Hence, 99.97% of the variation in the response Y in our sample can be explained by the linear model.

12.33 $f = 5.11$ with P-value $= 0.0303$. At level of 0.01, we fail to reject H_0 and we cannot claim that the regression is significant.

12.35 $f = \frac{(6.90079 - 1.13811)/1}{0.16508} = 34.91$ with 1 and 9 degrees of freedom. The P-value $= 0.0002$ which implies that H_0 is rejected.

12.37 The hypotheses are:

$$H_0 : \beta_1 = \beta_2 = 0,$$

$$H_1 : \text{At least one of the } \beta_i\text{'s is not zero, for } i = 1, 2.$$

The partial f-test statistic is

$$f = \frac{(4957.24074 - 17.02338)/2}{242.71561} = 10.18, \quad \text{with 2 and 7 degrees of freedom.}$$

The resulting P-value $= 0.0085$. Therefore, we reject H_0 and claim that at least one of β_1 and β_2 is not zero.

12.39 The following is the summary.

	s^2	R^2	R^2_{adj}
The model using weight alone	8.12709	0.8155	0.8104
The model using weight and drive ratio	4.78022	0.8945	0.8885

The above favor the model using both explanatory variables. Furthermore, in the model with two independent variables, the t-test for β_2, the coefficient of drive ratio, shows P-value < 0.0001. Hence, the drive ratio variable is important.

12.41 The following is the summary:

	s^2	C.V.	R^2_{adj}
The model with 3 terms	0.41966	4.62867	0.9807
The model without 3 terms	1.60019	9.03847	0.9266

Furthermore, to test $\beta_{11} = \beta_{12} = \beta_{22} = 0$ using the full model, $f = 15.07$ with P-value $= 0.0002$. Hence, the model with interaction and pure quadratic terms is better.

12.43 The following is the summary:

	s^2	C.V.	R^2_{adj}	Average Length of the CIs
x_1, x_2	650.14075	16.55705	0.7696	106.60577
x_1	967.90773	20.20209	0.6571	94.31092
x_2	679.99655	16.93295	0.7591	78.81977

In addition, in the full model when the individual coefficients are tested, we obtain P-value $= 0.3562$ for testing $\beta_1 = 0$ and P-value $= 0.1841$ for testing $\beta_2 = 0$. In comparing the three models, it appears that the model with x_2 only is slightly better.

12.45 (a) $\hat{y} = 5.95931 - 0.00003773$ odometer $+ 0.33735$ octane $- 12.62656$ van $- 12.98455$ suv.

(b) Since the coefficients of *van* and *suv* are both negative, sedan should have the best gas mileage.

(c) The parameter estimates (standard errors) for *van* and *suv* are -12.63 (1.07) and $-12,98$ (1.11), respectively. So, the difference between the estimates are smaller than one standard error of each. So, no significant difference in a *van* and an *suv* in terms of gas mileage performance.

12.47 (a) $\widehat{Hang\ Time} = 1.10765 + 0.01370 \text{ LLS} + 0.00429 \text{ Power}$.

(b) $\widehat{Hang\ Time} = 1.10765 + (0.01370)(180) + (0.00429)(260) - 4.6900$.

(c) $4.4502 < \mu_{Hang\ Time\ |\ \text{LLS}=180,\ \text{Power}=260} < 4.9299$.

12.49 Using computer output, with $\alpha = 0.05$, x_4 was removed first, and then x_1. Neither x_2 nor x_3 were removed and the final model is $\hat{y} = 2.18332 + 0.95758x_2 + 3.32533x_3$.

12.51   (a) $\hat{y} = -587.21085 + 428.43313x$.

(b) $\hat{y} = 1180.00032 - 192.69121x + 35.20945x^2$.

(c) The summary of the two models are given as:

Model	s^2	R^2	PRESS
$\mu_Y = \beta_0 + \beta_1 x$	1,105,054	0.8378	18,811,057.08
$\mu_Y = \beta_0 + \beta_1 x + \beta_{11} x^2$	430,712	0.9421	8,706,973.57

It appears that the model with a quadratic term is preferable.

12.53 $\hat{\sigma}_{b_1}^2 = 20,588.038$, $\hat{\sigma}_{b_{11}}^2 = 62.650$, and $\hat{\sigma}_{b_1 b_{11}} = -1,103.423$.

12.55   (a) There are many models here so the model summary is not displayed. By using MSE criterion, the best model, contains variables x_1 and x_3 with $s^2 = 313.491$. If PRESS criterion is used, the best model contains only the constant term with $s^2 = 317.51$. When the C_p method is used, the best model is model with the constant term.

(b) The normal probability plot, for the model using intercept only, is shown next. We do not appear to have the normality.

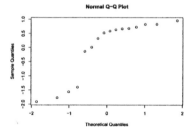

12.57   (a) $\hat{y} = 3.13682 + 0.64443x_1 - 0.01042x_2 + 0.50465x_3 - 0.11967x_4 - 2.46177x_5 + 1.50441x_6$.

(b) The final model using the stepwise regression is

$$\hat{y} = 4.65631 + 0.51133x_3 - 0.12418x_4.$$

(c) Using C_p criterion (smaller the better), the best model is still the model stated in (b) with $s^2 = 0.73173$ and $R^2 = 0.64758$. Using the s^2 criterion, the model with x_1, x_3 and x_4 has the smallest value of 0.72507 and $R^2 = 0.67262$. These two models are quite competitive. However, the model with two variables has one less variable, and thus may be more appealing.

(d) Using the model in part (b), displayed next is the Studentized residual plot. Note that observations 2 and 14 are beyond the 2 standard deviation lines. Both of those observations may need to be checked.

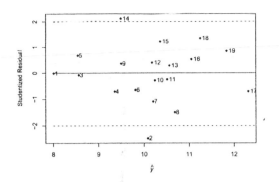

12.59 (a) $\hat{y} = 125.86555 + 7.75864x_1 + 0.09430x_2 - 0.00919x_1x_2$.

(b) The following is the summary of the models.

Model	s^2	R^2	PRESS	C_p
x_2	680.00	0.80726	7624.66	2.8460
x_1	967.91	0.72565	12310.33	4.8978
x_1x_2	650.14	0.86179	12696.66	3.4749
$x_1x_2x_3$	561.28	0.92045	15556.11	4.0000

It appears that the model with x_2 alone is the best.

Chapter 13

One-Factor Experiments: General

13.1 Using the formula of SSE, we have

$$SSE = \sum_{i=1}^{k}\sum_{j=1}^{n}(y_{ij} - \bar{y}_{i.})^2 = \sum_{i=1}^{k}\sum_{j=1}^{n}(\epsilon_{ij} - \bar{\epsilon}_{i.})^2 = \sum_{i=1}^{k}\left[\sum_{j=1}^{n}\epsilon_{ij}^2 - n\bar{\epsilon}_{i.}^2\right].$$

Hence

$$E(SSE) = \sum_{i=1}^{k}\left[\sum_{j=1}^{n}E(\epsilon_{ij}^2) - nE(\bar{\epsilon}_{i.}^2)\right] = \sum_{i=1}^{k}\left[n\sigma^2 - n\frac{\sigma^2}{n}\right] = k(n-1)\sigma^2.$$

Thus $E\left[\frac{SSE}{k(n-1)}\right] = \frac{k(n-1)\sigma^2}{k(n-1)} = \sigma^2.$

13.3 The hypotheses are

$$H_0 : \mu_1 = \mu_2 = \cdots = \mu_6,$$
$$H_1 : \text{At least two of the means are not equal.}$$

$\alpha = 0.05.$
Critical region: $f > 2.77$ with $v_1 = 5$ and $v_2 = 18$ degrees of freedom.
Computation:

Source of Variation	Sum of Squares	Degrees of Freedom	Mean Square	Computed f
Treatment	5.34	5	1.07	0.31
Error	62.64	18	3.48	
Total	67.98	23		

with P-value=0.9024.
Decision: The treatment means do not differ significantly.

13.5 The hypotheses are

$$H_0 : \mu_1 = \mu_2 = \mu_3,$$
$$H_1 : \text{At least two of the means are not equal.}$$

$\alpha = 0.01$.
Computation:

Source of Variation	Sum of Squares	Degrees of Freedom	Mean Square	Computed f
Shelf Height	399.3	2	199.63	14.52
Error	288.8	21	13.75	
Total	688.0	23		

with P-value=0.0001.
Decision: Reject H_0. The amount of money spent on dog food differs with the shelf height of the display.

13.7 The hypotheses are

$$H_0 : \mu_1 = \mu_2 = \mu_3 = \mu_4,$$
$$H_1 : \text{At least two of the means are not equal.}$$

$\alpha = 0.05$.
Computation:

Source of Variation	Sum of Squares	Degrees of Freedom	Mean Square	Computed f
Treatments	119.787	3	39.929	2.25
Error	638.248	36	17.729	
Total	758.035	39		

with P-value=0.0989.
Decision: Fail to reject H_0 at level $\alpha = 0.05$.

13.9 The hypotheses are

$$H_0 : \mu_1 = \mu_2 = \mu_3 = \mu_4,$$
$$H_1 : \text{At least two of the means are not equal.}$$

$\alpha = 0.01$.
Computation:

Source of Variation	Sum of Squares	Degrees of Freedom	Mean Square	Computed f
Treatments	27.5506	3	9.1835	8.38
Error	18.6360	17	1.0962	
Total	46.1865	20		

with P-value= 0.0012.

Decision: Reject H_0. Average specific activities differ.

13.11 Computation:

Source of Variation	Sum of Squares	Degrees of Freedom	Mean Square	Computed f
B vs. A, C, D	30.6735	1	30.6735	14.28
C vs. A, D	49.9230	1	49.9230	23.23
A vs. D	5.3290	1	5.3290	2.48
Error	34.3800	16	2.1488	

(a) P-value=0.0016. B is significantly different from the average of A, C, and D.

(b) P-value=0.0002. C is significantly different from the average of A and D.

(c) P-value=0.1349. A can not be shown to differ significantly from D.

13.13 (a) The hypotheses are

$$H_0 : \mu_1 = \mu_2 = \mu_3 = \mu_4,$$
$$H_1 : \text{At least two of the means are not equal.}$$

Source of Variation	Sum of Squares	Degrees of Freedom	Mean Square	Computed f
Treatments	1083.60	3	361.20	13.50
Error	1177.68	44	26.77	
Total	2261.28	47		

with P-value< 0.0001.

Decision: Reject H_0. The treatment means are different.

(b) For testing two contrasts $L_1 = \mu_1 - \mu_2$ and $L_2 = \mu_3 - \mu_4$ at level $\alpha = 0.01$, we have the following

Contrast	Sum of Squares	Computed f	P-value
1 vs. 2	785.47	29.35	< 0.0001
3 vs. 4	96.00	3.59	0.0648

Hence, Bath I and Bath II were significantly different for 5 launderings, and Bath I and Bath II were not different for 10 launderings.

13.15 Since $q(0.05, 4, 16) = 4.05$, the critical difference is $(4.05)\sqrt{\frac{2.14875}{5}} = 2.655$. Hence

$\bar{y}_{3.}$	$\bar{y}_{1.}$	$\bar{y}_{4.}$	$\bar{y}_{2.}$
56.52	59.66	61.12	61.96

13.17   (a) The hypotheses are

$$H_0 : \mu_1 = \mu_2 = \cdots = \mu_5,$$
$$H_1 : \text{At least two of the means are not equal.}$$

Computation:

Source of Variation	Sum of Squares	Degrees of Freedom	Mean Square	Computed f
Procedures	7828.30	4	1957.08	9.01
Error	3256.50	15	217.10	
Total	11084.80	19		

with P-value= 0.0006.

Decision: Reject H_0. There is a significant difference in the average species count for the different procedures.

(b) Since $q(0.05, 5, 15) = 4.373$ and $\sqrt{\frac{217.10}{4}} = 7.367$, the critical difference is 32.2. Hence

$\bar{y}_K$	$\bar{y}_S$	$\bar{y}_{Sub}$	$\bar{y}_M$	$\bar{y}_D$
12.50	24.25	26.50	55.50	64.25

13.19 When we obtain the ANOVA table, we derive $s^2 = 0.2174$. Hence

$$\sqrt{2s^2/n} = \sqrt{(2)(0.2174)/5} = 0.2949.$$

The sample means for each treatment levels are

$$\bar{y}_C = 6.88, \quad \bar{y}_{1.} = 8.82, \quad \bar{y}_{2.} = 8.16, \quad \bar{y}_{3.} = 6.82, \quad \bar{y}_{4.} = 6.14.$$

Hence

$$d_1 = \frac{8.82 - 6.88}{0.2949} = 6.579, \quad d_2 = \frac{8.16 - 6.88}{0.2949} = 4.340,$$
$$d_3 = \frac{6.82 - 6.88}{0.2949} = -0.203, \quad d_4 = \frac{6.14 - 6.88}{0.2949} = -2.509.$$

From Table A.14, we have $d_{0.025}(4, 20) = 2.65$. Therefore, concentrations 1 and 2 are significantly different from the control.

13.21 Aggregate 4 has a significantly lower absorption rate than the other aggregates.

13.23 The ANOVA table can be obtained as follows:

Source of Variation	Sum of Squares	Degrees of Freedom	Mean Square	Computed f
Temperatures	1268.5333	4	317.1333	70.27
Error	112.8333	25	4.5133	
Total	1381.3667	29		

with P-value< 0.0001.

The results from Tukey's procedure can be obtained as follows:

$$\begin{array}{ccccc} \bar{y}_0 & \bar{y}_{25} & \bar{y}_{100} & \bar{y}_{75} & \bar{y}_{50} \\ 55.167 & 60.167 & 64.167 & 70.500 & 72.833 \end{array}$$

The batteries activated at temperature 50 and 75 have significantly longer activated life.

13.25 Based on the definition, we have the following.

$$SSB = k\sum_{j=1}^{b}(\bar{y}_{.j} - \bar{y}_{..})^2 = k\sum_{j=1}^{b}\left(\frac{T_{.j}}{k} - \frac{T_{..}}{bk}\right)^2 = \sum_{j=1}^{b}\frac{T_{.j}^2}{k} - 2\frac{T_{..}^2}{bk} + \frac{T_{..}^2}{bk} = \sum_{j=1}^{b}\frac{T_{.j}^2}{k} - \frac{T_{..}^2}{bk}.$$

13.27 (a) The hypotheses are

$$H_0 : \alpha_1 = \alpha_2 = \alpha_3 = \alpha_4 = 0, \text{ fertilizer effects are zero}$$

$$H_1 : \text{At least one of the } \alpha_i\text{'s is not equal to zero.}$$

$\alpha = 0.05$.

Critical region: $f > 4.76$.

Computation:

Source of Variation	Sum of Squares	Degrees of Freedom	Mean Square	Computed f
Fertilizers	218.1933	3	72.7311	6.11
Blocks	197.6317	2	98.8158	
Error	71.4017	6	11.9003	
Total	487.2267	11		

P-value$= 0.0296$. Decision: Reject H_0. The means are not all equal.

(b) The results of testing the contrasts are shown as:

Source of Variation	Sum of Squares	Degrees of Freedom	Mean Square	Computed f
(f_1, f_3) vs (f_2, f_4)	206.6700	1	206.6700	17.37
f_1 vs f_3	11.4817	1	11.4817	0.96
Error	71.4017	6	11.9003	

The corresponding P-values for the above contrast tests are 0.0059 and 0.3639, respectively. Hence, for the first contrast, the test is significant and the for the second contrast, the test is insignificant.

13.29 The hypotheses are

$$H_0 : \alpha_1 = \alpha_2 = \alpha_3 = 0, \text{ brand effects are zero}$$

$$H_1 : \text{At least one of the } \alpha_i\text{'s is not equal to zero.}$$

$\alpha = 0.05$.
Critical region: $f > 3.84$.
Computation:

Source of Variation	Sum of Squares	Degrees of Freedom	Mean Square	Computed f
Treatments	27.797	2	13.899	5.99
Blocks	16.536	4	4.134	
Error	18.556	8	2.320	
Total	62.889	14		

P-value=0.0257. Decision: Reject H_0; mean percent of foreign additives is not the same for all three brand of jam. The means are:

Jam A: 2.36,   Jam B: 3.48,   Jam C: 5.64.

Based on the means, Jam A appears to have the smallest amount of foreign additives.

13.31 The hypotheses are

$$H_0 : \alpha_1 = \alpha_2 = \cdots = \alpha_6 = 0, \text{ station effects are zero}$$

$$H_1 : \text{At least one of the } \alpha_i\text{'s is not equal to zero.}$$

$\alpha = 0.01$.
Computation:

Source of Variation	Sum of Squares	Degrees of Freedom	Mean Square	Computed f
Stations	230.127	5	46.025	26.14
Dates	3.259	5	0.652	
Error	44.018	25	1.761	
Total	277.405	35		

P-value< 0.0001. Decision: Reject H_0; the mean concentration is different at the different stations.

13.33 The hypotheses are

$$H_0 : \alpha_1 = \alpha_2 = \alpha_3 = 0, \text{ diet effects are zero}$$

$$H_1 : \text{At least one of the } \alpha_i \text{'s is not equal to zero.}$$

$\alpha = 0.01$.
Computation:

Source of Variation	Sum of Squares	Degrees of Freedom	Mean Square	Computed f
Diets	4297.000	2	2148.500	11.86
Subjects	6033.333	5	1206.667	
Error	1811.667	10	181.167	
Total	12142.000	17		

P-value= 0.0023. Decision: Reject H_0; differences among the diets are significant.

13.35 The hypotheses are

$$H_0 : \alpha_1 = \alpha_2 = \alpha_3 = \alpha_4 = \alpha_5 = 0, \text{ treatment effects are zero}$$

$$H_1 : \text{At least one of the } \alpha_i \text{'s is not equal to zero.}$$

$\alpha = 0.01$.
Computation:

Source of Variation	Sum of Squares	Degrees of Freedom	Mean Square	Computed f
Treatments	79630.133	4	19907.533	0.58
Locations	634334.667	5	126866.933	
Error	689106.667	20	34455.333	
Total	1403071.467	29		

P-value= 0.6821. Decision: Do not reject H_0; the treatment means do not differ significantly.

13.37 The total sums of squares can be written as

$$
\begin{aligned}
\sum_i \sum_j \sum_k (y_{ijk} - \bar{y}_{...})^2 &= \sum_i \sum_j \sum_k [(\bar{y}_{i..} - \bar{y}_{...}) + (\bar{y}_{.j.} - \bar{y}_{...}) + (\bar{y}_{..k} - \bar{y}_{...}) \\
&\quad + (y_{ijk} - \bar{y}_{i..} - \bar{y}_{.j.} - \bar{y}_{..k} + 2\bar{y}_{...})]^2 \\
&= r \sum_i (\bar{y}_{i..} - \bar{y}_{...})^2 + r \sum_j (\bar{y}_{.j.} - \bar{y}_{...})^2 + r \sum_k (\bar{y}_{..k} - \bar{y}_{...})^2 \\
&\quad + \sum_i \sum_j \sum_k (y_{ijk} - \bar{y}_{i..} - \bar{y}_{.j.} - \bar{y}_{..k} + 2\bar{y}_{...})^2 \\
&\quad + 6 \text{ cross-product terms,}
\end{aligned}
$$

and all cross-product terms are equal to zeroes.

13.39 The hypotheses are

$$H_0 : \tau_1 = \tau_2 = \tau_3 = \tau_4 = 0, \text{ professor effects are zero}$$

$$H_1 : \text{At least one of the } \tau_i\text{'s is not equal to zero.}$$

$\alpha = 0.05$.
Computation:

Source of Variation	Sum of Squares	Degrees of Freedom	Mean Square	Computed f
Time Periods	474.50	3	158.17	
Courses	252.50	3	84.17	
Professors	723.50	3	241.17	5.03
Error	287.50	6	47.92	
Total	1738.00	15		

P-value= 0.0446. Decision: Reject H_0; grades are affected by different professors.

13.41 The hypotheses are

$$H_0 : \alpha_1 = \alpha_2 = \alpha_3 = 0, \text{ dye effects are zero}$$

$$H_1 : \text{At least one of the } \alpha_i\text{'s is not equal to zero.}$$

$\alpha = 0.05$.
Computation:

Source of Variation	Sum of Squares	Degrees of Freedom	Mean Square	Computed f
Amounts	1238.8825	2	619.4413	122.37
Plants	53.7004	1	53.7004	
Error	101.2433	20	5.0622	
Total	1393.8263	23		

P-value< 0.0001. Decision: Reject H_0; color densities of fabric differ significantly for three levels of dyes.

13.43 (a) The hypotheses are

$$H_0 : \sigma_\alpha^2 = 0,$$
$$H_1 : \sigma_\alpha^2 \neq 0$$

$\alpha = 0.05$.
Computation:

Source of Variation	Sum of Squares	Degrees of Freedom	Mean Square	Computed f
Operators	371.8719	3	123.9573	14.91
Error	99.7925	12	8.3160	
Total	471.6644	15		

P-value= 0.0002. Decision: Reject H_0; operators are different.

(b) $\hat{\sigma}^2 = 8.316$ and $\hat{\sigma}_\alpha^2 = \frac{123.9573 - 8.3160}{4} = 28.910$.

13.45 (a) The hypotheses are

$$H_0 : \sigma_\alpha^2 = 0,$$
$$H_1 : \sigma_\alpha^2 \neq 0$$

$\alpha = 0.05$.

Computation:

Source of Variation	Sum of Squares	Degrees of Freedom	Mean Square	Computed f
Treatments	23.238	3	7.746	3.33
Blocks	45.283	4	11.321	
Error	27.937	12	2.328	
Total	96.458	19		

P-value= 0.0565. Decision: Not able to show a significant difference in the random treatments at 0.05 level, although the P-value shows marginal significance.

(b) $\sigma_\alpha^2 = \frac{7.746 - 2.328}{5} = 1.084$, and $\sigma_\beta^2 = \frac{11.321 - 2.328}{4} = 2.248$.

13.47 (a) The matrix is

$$\mathbf{A} = \begin{bmatrix} bk & b & b & \cdots & b & k & k & \cdots & k \\ b & b & 0 & \cdots & 0 & 1 & 1 & \cdots & 1 \\ b & 0 & b & \cdots & 0 & 1 & 1 & \cdots & 1 \\ \vdots & \vdots & \vdots & \ddots & \vdots & \vdots & \vdots & \ddots & \vdots \\ b & 0 & 0 & \cdots & b & 1 & 1 & \cdots & 1 \\ k & 1 & 1 & \cdots & k & 0 & 0 & \cdots & 0 \\ k & 1 & 1 & \cdots & 0 & k & 0 & \cdots & 0 \\ \vdots & \vdots & \vdots & \ddots & \vdots & \vdots & \vdots & \ddots & \vdots \\ k & 1 & 1 & \cdots & 0 & 0 & 0 & \cdots & k \end{bmatrix},$$

where b = number of blocks and k = number of treatments. The vectors are

$$\mathbf{b}' = (\mu, \alpha_1, \alpha_2, \cdots, \alpha_k, \beta_1, \beta_2, \cdots, \beta_b)', \quad \text{and}$$
$$\mathbf{g}' = (T_{..}, T_{1.}, T_{2.}, \cdots, T_{k.}, T_{.1}, T_{.2}, \cdots, T_{.b})'.$$

(b) Solving the system $\mathbf{Ab} = \mathbf{g}$ with the constraints $\sum\limits_{i=1}^{k} \alpha_i = 0$ and $\sum\limits_{j=1}^{b} \beta_j = 0$, we have

$$\hat{\mu} = \bar{y}_{..},$$

$$\hat{\alpha}_i = \bar{y}_{i.} - \bar{y}_{..}, \quad \text{for } i = 1, 2, \ldots, k,$$

$$\hat{\beta}_j = \bar{y}_{.j} - \bar{y}_{..}, \quad \text{for } j = 1, 2, \ldots, b.$$

Therefore,

$$R(\alpha_1, \alpha_2, \ldots, \alpha_k, \beta_1, \beta_2, \ldots, \beta_b) = \mathbf{b}'\mathbf{g} - \frac{T_{..}^2}{bk}$$

$$= \sum_{i=1}^{k} \frac{T_{i.}^2}{b} + \sum_{j=1}^{b} \frac{T_{.j}^2}{k} - 2\frac{T_{..}^2}{bk}.$$

To find $R(\beta_1, \beta_2, \ldots, \beta_b \mid \alpha_1, \alpha_2, \ldots, \alpha_k)$ we first find $R(\alpha_1, \alpha_2, \ldots, \alpha_k)$. Setting $\beta_j = 0$ in the model, we obtain the estimates (after applying the constraint $\sum\limits_{i=1}^{k} \alpha_i = 0$)

$$\hat{\mu} = \bar{y}_{..}, \quad \text{and} \quad \hat{\alpha}_i = \bar{y}_{i.} - \bar{y}_{..}, \quad \text{for } i = 1, 2, \ldots, k.$$

The $\mathbf{g}$ vector is the same as in part (a) with the exception that $T_{.1}, T_{.2}, \ldots, T_{.b}$ do not appear. Thus one obtains

$$R(\alpha_1, \alpha_2, \ldots, \alpha_k) = \sum_{i=1}^{k} \frac{T_{i.}^2}{b} - \frac{T_{..}^2}{bk}$$

and thus

$$R(\beta_1, \beta_2, \ldots, \beta_b \mid \alpha_1, \alpha_2, \ldots, \alpha_k) = R(\alpha_1, \alpha_2, \ldots, \alpha_k, \beta_1, \beta_2, \ldots, \beta_b)$$

$$- R(\alpha_1, \alpha_2, \ldots, \alpha_k) = \sum_{j=1}^{b} \frac{T_{.j}^2}{k} - \frac{T_{..}^2}{bk} = SSB.$$

13.49 We know $\phi^2 = b \sum\limits_{i=1}^{4} \frac{\alpha_i^2}{4\sigma^2} = \frac{b}{2}$, when $\sum\limits_{i=1}^{4} \frac{\alpha_i^2}{\sigma^2} = 2.0$.

If $b = 10$, $\phi = 2.24$; $v_1 = 3$ and $v_2 = 27$ degrees of freedom.
If $b = 9$, $\phi = 2.12$; $v_1 = 3$ and $v_2 = 24$ degrees of freedom.
If $b = 8$, $\phi = 2.00$; $v_1 = 3$ and $v_2 = 21$ degrees of freedom.
From Table A.16 we see that $b = 9$ gives the desired result.

13.51 (a) The model is $y_{ij} = \mu + \alpha_i + \epsilon_{ij}$, where $\alpha_i \sim n(0, \sigma_\alpha^2)$.

(b) Since $s^2 = 0.02056$ and $s_1^2 = 0.01791$, we have $\hat{\sigma}^2 = 0.02056$ and $\frac{s_1^2 - s^2}{10} = \frac{0.01791 - 0.02056}{10} = -0.00027$, which implies $\hat{\sigma}_\alpha^2 = 0$.

13.53 (a) $y_{ij} = \mu + \alpha_i + \epsilon_{ij}$, where $\alpha_i \sim n(x; 0, \sigma_\alpha^2)$.

(b) Running an ANOVA analysis, we obtain the P-value as 0.0121. Hence, the loom variance component is significantly different from 0 at level 0.05.

(c) The suspicion is supported by the data.

Chapter 14

Factorial Experiments (Two or More Factors)

14.1 The hypotheses of the three parts are,

(a) for the main effects temperature,

$$H_0' : \alpha_1 = \alpha_2 = \alpha_3 = 0,$$

$$H_1' : \text{At least one of the } \alpha_i\text{'s is not zero};$$

(b) for the main effects ovens,

$$H_0'' : \beta_1 = \beta_2 = \beta_3 = \beta_4 = 0,$$

$$H_1'' : \text{At least one of the } \beta_i\text{'s is not zero};$$

(c) and for the interactions,

$$H_0''' : (\alpha\beta)_{11} = (\alpha\beta)_{12} = \cdots = (\alpha\beta)_{34} = 0,$$

$$H_1''' : \text{At least one of the } (\alpha\beta)_{ij}\text{'s is not zero}.$$

$\alpha = 0.05$.

Critical regions: (a) $f_1 > 3.00$; (b) $f_2 > 3.89$; and (c) $f_3 > 3.49$.

Computations: From the computer printout we have

Source of Variation	Sum of Squares	Degrees of Freedom	Mean Square	Computed f
Temperatures	5194.08	2	2597.0400	8.13
Ovens	4963.12	3	1654.3733	5.18
Interaction	3126.26	6	521.0433	1.63
Error	3833.50	12	319.4583	
Total	17,116.96	23		

Decision: (a) Reject H_0'; (b) Reject H_0''; (c) Do not reject H_0'''.

14.3 The hypotheses of the three parts are,

   (a) for the main effects environments,

$$H_0' : \alpha_1 = \alpha_2 = 0, \text{ (no differences in the environment)}$$

$$H_1' : \text{At least one of the } \alpha_i\text{'s is not zero;}$$

   (b) for the main effects strains,

$$H_0'' : \beta_1 = \beta_2 = \beta_3 = 0, \text{ (no differences in the strains)}$$

$$H_1'' : \text{At least one of the } \beta_i\text{'s is not zero;}$$

   (c) and for the interactions,

$$H_0''' : (\alpha\beta)_{11} = (\alpha\beta)_{12} = \cdots = (\alpha\beta)_{23} = 0, \text{ (environments and strains do not interact)}$$

$$H_1''' : \text{At least one of the } (\alpha\beta)_{ij}\text{'s is not zero.}$$

$\alpha = 0.01$.
Critical regions: (a) $f_1 > 7.29$; (b) $f_2 > 5.16$; and (c) $f_3 > 5.16$.
Computations: From the computer printout we have

Source of Variation	Sum of Squares	Degrees of Freedom	Mean Square	Computed f
Environments	14,875.521	1	14,875.521	14.81
Strains	18,154.167	2	9,077.083	9.04
Interaction	1,235.167	2	617.583	0.61
Error	42,192.625	42	1004.586	
Total	76,457.479	47		

Decision: (a) Reject H_0'; (b) Reject H_0''; (c) Do not reject H_0'''. Interaction is not significant, while both main effects, environment and strain, are all significant.

14.5 The hypotheses of the three parts are,

   (a) for the main effects subjects,

$$H_0' : \alpha_1 = \alpha_2 = \alpha_3 = 0,$$

$$H_1' : \text{At least one of the } \alpha_i\text{'s is not zero;}$$

   (b) for the main effects muscles,

$$H_0'' : \beta_1 = \beta_2 = \beta_3 = \beta_4 = \beta_5 = 0,$$

$$H_1'' : \text{At least one of the } \beta_i\text{'s is not zero;}$$

(c) and for the interactions,

$$H_0''' : (\alpha\beta)_{11} = (\alpha\beta)_{12} = \cdots = (\alpha\beta)_{35} = 0,$$

$$H_1''' : \text{At least one of the } (\alpha\beta)_{ij}\text{'s is not zero.}$$

$\alpha = 0.01$.

Critical regions: (a) $f_1 > 5.39$; (b) $f_2 > 4.02$; and (c) $f_3 > 3.17$.

Computations: From the computer printout we have

Source of Variation	Sum of Squares	Degrees of Freedom	Mean Square	Computed f
Subjects	4,814.74	2	2,407.37	34.40
Muscles	7,543.87	4	1,885.97	26.95
Interaction	11,362.20	8	1,420.28	20.30
Error	2,099.17	30	69.97	
Total	25,819.98	44		

Decision: (a) Reject H_0'; (b) Reject H_0''; (c) Reject H_0'''.

14.7 The ANOVA table is

Source of Variation	Sum of Squares	Degrees of Freedom	Mean Square	Computed f	P-value
Temperature	430.475	3	143.492	10.85	0.0003
Catalyst	2,466.650	4	616.663	46.63	< 0.0001
Interaction	326.150	12	27.179	2.06	0.0745
Error	264.500	20	13.225		
Total	3,487.775	39			

Decision: All main effects are significant and the interaction is significant at level 0.0745. Hence, if 0.05 significance level is used, interaction is not significant. An interaction plot is given here.

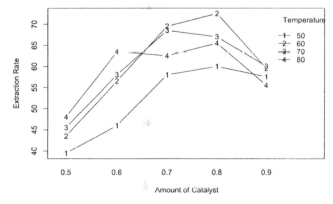

Duncan's tests, at level 0.05, for both main effects result in the following.

(a) For Temperature:

$$\begin{array}{cccc} \bar{y}_{50} & \bar{y}_{80} & \bar{y}_{70} & \bar{y}_{60} \\ \underline{52.200} & \underline{59.000} & \underline{59.800} & \underline{60.300} \end{array}$$

(b) For Amount of Catalyst:

$$\begin{array}{ccccc} \bar{y}_{0.5} & \bar{y}_{0.6} & \bar{y}_{0.9} & \bar{y}_{0.7} & \bar{y}_{0.8} \\ \underline{44.125} & \underline{56.000} & \underline{58.125} & \underline{64.625} & \underline{66.250} \end{array}$$

14.9 (a) The ANOVA table is

Source of Variation	Sum of Squares	Degrees of Freedom	Mean Square	Computed f	P-value
Tool	675.00	1	675.00	74.31	< 0.0001
Speed	12.00	1	12.00	1.32	0.2836
Tool*Speed	192.00	1	192.00	21.14	0.0018
Error	72.67	8	9.08		
Total	951.67	11			

Decision: The interaction effects are significant. Although the main effects of speed showed insignificance, we might not make such a conclusion since its effects might be masked by significant interaction.

(b) In the graph shown, we claim that the cutting speed that results in the longest life of the machine tool depends on the tool geometry, although the variability of the life is greater with tool geometry at level 1.

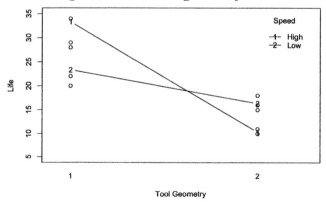

(c) Since interaction effects are significant, we do the analysis of variance for separate tool geometry.

   (i) For tool geometry 1, an f-test for the cutting speed resulted in a P-value = 0.0405 with the mean life (standard deviation) of the machine tool at 33.33 (4.04) for high speed and 23.33 (4.16) for low speed. Hence, a high cutting speed has longer life for tool geometry 1.

   (ii) For tool geometry 2, an f-test for the cutting speed resulted in a P-value = 0.0031 with the mean life (standard deviation) of the machine tool at 10.33

(0.58) for high speed and 16.33 (1.53) for low speed. Hence, a low cutting speed has longer life for tool geometry 2.

For the above detailed analysis, we note that the standard deviations for the mean life are much higher at tool geometry 1.

(d) See part (b).

14.11 (a) The ANOVA table is

Source of Variation	Sum of Squares	Degrees of Freedom	Mean Square	Computed f	P-value
Method	0.000104	1	0.000104	6.57	0.0226
Lab	0.008058	6	0.001343	84.70	< 0.0001
Method*Lab	0.000198	6	0.000033	2.08	0.1215
Error	0.000222	14	0.000016		
Total	0.00858243	27			

(b) Since the P-value $= 0.1215$ for the interaction, the interaction is not significant. Hence, the results on the main effects can be considered meaningful to the scientist.

(c) Both main effects, method of analysis and laboratory, are all significant.

(d) The interaction plot is show here.

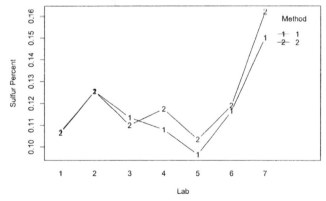

(e) When the tests are done separately, i.e., we only use the data for Lab 1, or Lab 2 alone, the P-values for testing the differences of the methods at Lab 1 and 7 are 0.8600 and 0.1557, respectively. In this case, usually the degrees of freedom of errors are small. If we compare the mean differences of the method within the overall ANOVA model, we obtain the P-values for testing the differences of the methods at Lab 1 and 7 as 0.9010 and 0.0093, respectively. Hence, methods are no difference in Lab 1 and are significantly different in Lab 7. Similar results may be found in the interaction plot in (d).

14.13 (a) The interaction plot is show here. There seems no interaction effect.

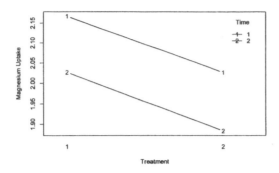

(b) The ANOVA table is

Source of Variation	Sum of Squares	Degrees of Freedom	Mean Square	Computed f	P-value
Treatment	0.060208	1	0.060208	157.07	< 0.0001
Time	0.060208	1	0.060203	157.07	< 0.0001
Treatment*Time	0.000008	1	0.000008	0.02	0.8864
Error	0.003067	8	0.000383		
Total	0.123492	11			

(c) The magnesium uptake are lower using treatment 2 than using treatment 1, no matter what the times are. Also, time 2 has lower magnesium uptake than time 1. All the main effects are significant.

(d) Using the regression model and making "Treatment" as categorical, we have the following fitted model:

$$\hat{y} = 2.4433 - 0.13667 \text{ Treatment} - 0.13667 \text{ Time} - 0.00333 \text{Treatment} \times \text{Time}.$$

(e) The P-value of the interaction for the above regression model is 0.8864 and hence it is insignificant.

14.15 The ANOVA table is given here.

Source of Variation	Sum of Squares	Degrees of Freedom	Mean Square	Computed f	P-value
Main effect					
A	2.24074	1	2.24074	0.54	0.4652
B	56.31815	2	28.15907	6.85	0.0030
C	17.65148	2	8.82574	3.83	0.1316
Two-factor Interaction					
AB	31.47148	2	15.73574	3.83	0.0311
AC	31.20259	2	15.60130	3.79	0.0320
BC	2156074	4	5.39019	1.31	0.2845
Three-factor Interaction					
ABC	26.79852	4	6.69963	1.63	0.1881
Error	148.04000	36	4.11221		
Total	335.28370	53			

(a) Based on the P-values, only AB and AC interactions are significant.

(b) The main effect B is significant. However, due to significant interactions mentioned in (a), the insignificance of A and C cannot be counted.

(c) Look at the interaction plot of the mean responses versus C for different cases of A.

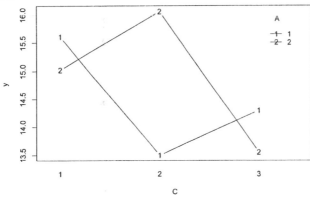

Apparently, the mean responses at different levels of C varies in different patterns for the different levels of A. Hence, although the overall test on factor C is insignificant, it is misleading since the significance of the effect C is masked by the significant interaction between A and C.

14.17 Letting A, B, and C designate coating, humidity, and stress, respectively, the ANOVA table is given here.

Source of Variation	Sum of Squares	Degrees of Freedom	Mean Square	Computed f	P-value
Main effect					
A	216, 384.1	1	216, 384.1	0.05	0.8299
B	19, 876, 891.0	2	9, 938, 445.5	2.13	0.1257
C	427, 993, 946.4	2	213, 996, 973.2	45.96	< 0.0001
Two-factor Interaction					
AB	31, 736, 625	2	15, 868, 312.9	3.41	0.0385
AC	699, 830.1	2	349, 915.0	0.08	0.9277
BC	58, 623, 693.2	4	13, 655, 923.3	3.15	0.0192
Three-factor Interaction					
ABC	36, 034, 808.9	4	9, 008, 702.2	1.93	0.1138
Error	335, 213, 133.6	72	4, 655, 738.0		
Total	910, 395, 313.1	89			

(a) The Coating and Humidity interaction, and the Humidity and Stress interaction have the *P*-values of 0.0385 and 0.0192, respectively. Hence, they are all significant. On the other hand, the Stress main effect is strongly significant as well. However, both other main effects, Coating and Humidity, cannot be claimed as insignificant, since they are all part of the two significant interactions.

(b) A Stress level of 20 consistently produces low fatigue. It appears to work best with medium humidity and an uncoated surface.

14.19 The ANOVA table shows:

Source of Variation	Sum of Squares	Degrees of Freedom	Mean Square	Computed f	P-value
A	0.16617	2	0.08308	14.22	< 0.0001
B	0.07825	2	0.03913	6.70	0.0020
C	0.01947	2	0.00973	1.67	0.1954
AB	0.12845	4	0.03211	5.50	0.0006
AC	0.06280	4	0.01570	2.69	0.0369
BC	0.12644	4	0.03161	5.41	0.0007
ABC	0.14224	8	0.01765	3.02	0.0051
Error	0.47323	81	0.00584		
Total	1.19603	107			

There is a significant three-way interaction by Temperature, Surface, and Hrc. A plot of each Temperature is given to illustrate the interaction

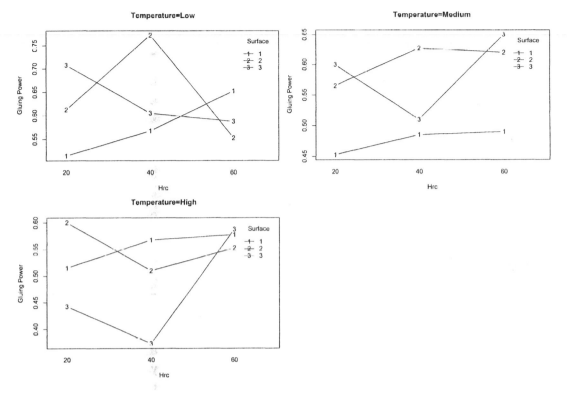

14.21 (a) Yes, the P-values for $Brand * Type$ and $Brand * Temp$ are both < 0.0001.

(b) The main effect of Brand has a P-value < 0.0001. So, three brands averaged across the other two factors are significantly different.

(c) Using brand Y, powdered detergent and hot water yields the highest percent removal of dirt.

14.23 (a) The P-values of two-way interactions Time×Temperature, Time×Solvent, Temperature × Solvent, and the P-value of the three-way interaction Time×Temperature×Solvent are 0.1103, 0.1723, 0.8558, and 0.0140, respectively.

(b) The interaction plots for different levels of Solvent are given here.

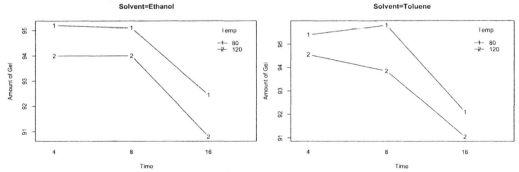

(c) A normal probability plot of the residuals is given and it appears that normality assumption may not be valid.

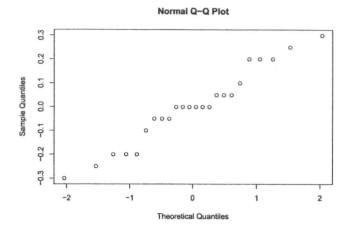

14.25 The ANOVA table is given.

Source of Variation	Sum of Squares	Degrees of Freedom	Mean Square	Computed f	P-value
Filters	4.63389	2	2.31694	8.39	0.0183
Operators	10.31778	3	3.43926	12.45	0.0055
Interaction	1.65722	6	0.27620	1.49	0.2229
Error	4.44000	24	0.18500		
Total	21.04889	35			

Note the f values for the main effects are using the interaction term as the denominator.

(a) The hypotheses are

$$H_0 : \sigma^2_{\alpha\beta} = 0,$$
$$H_1 : \sigma^2_{\alpha\beta} \neq 0.$$

Decision: Since P-value = 0.2229, the null hypothesis cannot be rejected. There is no significant interaction variance component.

(b) The hypotheses are

$$H_0' : \sigma^2_\alpha = 0. \qquad H_0'' : \sigma^2_\beta = 0.$$
$$H_1' : \sigma^2_\alpha \neq 0. \qquad H_1'' : \sigma^2_\beta \neq 0.$$

Decisions: Based on the P-values of 0.0183, and 0.0055 for H_0' and H_1', respectively, we reject both H_0' and H_0''. Both σ^2_α and σ^2_β are significantly different from zero.

(c) $\hat{\sigma}^2 = s^2 = 0.185$; $\hat{\sigma}^2_\alpha = \frac{2.31691 - 0.185}{12} = 0.17766$, and $\hat{\sigma}^2_\beta = \frac{3.43926 - 0.185}{9} = 0.35158.$

14.27 The ANOVA table with expected mean squares is given here.

Source of Variation	Degrees of Freedom	Mean Square	(a) Computed f	(b) Computed f
A	3	140	$f_1 = s_1^2/s_5^2 = 5.83$	$f_1 = s_1^2/s_5^2 = 5.83$
B	1	480	$f_2 = s_2^2/s_{p_1}^2 = 78.82$	$f_2 = s_2^2/s_6^2 = 26.67$
C	2	325	$f_3 = s_3^2/s_5^2 = 13.54$	$f_3 = s_3^2/s_5^2 = 13.54$
AB	3	15	$f_4 = s_4^2/s_{p_2}^2 = 2.86$	$f_4 = s_4^2/s_7^2 = 7.50$
AC	6	24	$f_5 = s_5^2/s_{p_2}^2 = 4.57$	$f_5 = s_5^2/s_7^2 = 12.00$
BC	2	18	$f_6 = s_6^2/s_{p_1}^2 = 4.09$	$f_6 = s_6^2/s_7^2 = 9.00$
ABC	6	2	$f_7 = s_7^2/s^2 = 0.40$	$f_7 = s_7^2/s^2 = 0.40$
Error	24	5		
Total	47			

In column (a) we have found the following main effects and interaction effects significant using the pooled estimates: σ_β^2, σ_γ^2, and $\sigma_{\alpha\gamma}^2$.

$s_{p_1}^2 = (12 + 120)/30 = 4.4$ with 30 degrees of freedom.

$s_{p_2}^2 = (12 + 120 + 36)/32 = 5.25$ with 32 degrees of freedom.

$s_{p_3}^2 = (12 + 120 + 36 + 45)/35 - 6.09$ with 35 degrees of freedom.

In column (b) we have found the following main effect and interaction effect significant when sums of squares of insignificant effects were not pooled: σ_γ^2 and $\sigma_{\alpha\gamma}^2$.

14.29 The power can be calculated as

$$1 - \beta = P\left[F(2,6) > f_{0.05}(2,6)\frac{\sigma^2 + 3\sigma_{\alpha\beta}^2}{\sigma^2 + 3\sigma_{\alpha\beta}^2 + 12\sigma_\beta^2}\right]$$

$$= P\left[F(2,6) > \frac{(5.14)(0.2762)}{2.3169}\right] = P[F(2,6) > 0.6127] = 0.57.$$

14.31 (a) A mixed model.

(b) The ANOVA table is

Source of Variation	Sum of Squares	Degrees of Freedom	Mean Square	Computed f	P-value
Material	1.03488	2	0.51744	47.42	< 0.0001
Brand	0.60654	2	0.30327	1.73	0.2875
Material*Brand	9,70109	4	0.17527	16.06	0.0004
Error	0.09820	9	0.01091		
Total	2.44071	17			

(c) No, the main effect of Brand is not significant. An interaction plot is given.

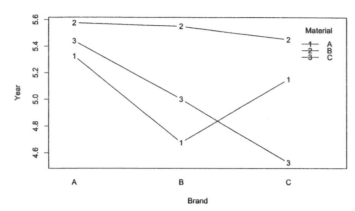

Although brand A has highest means in general, it is not always significant, especially for Material 2.

14.33 (a) A mixed model. Power setting (α_i in the model) is a fixed effect. Cereal type (β_j in the model) is a random effect.

$$y_{ijk} = \mu + \alpha_i + \beta_j + (\alpha\beta)_{ij} + \epsilon_{ijk};$$
$$\sum_i \alpha_i = 0, \quad \beta_j \sim n(x; 0, \sigma_\beta^2), \quad (\alpha\beta)_{ij} \sim n(x; 0, \sigma_{\alpha\beta}^2), \quad \epsilon_{ijk} \sim n(x; 0, \sigma^2).$$

(b) No. $f_{2,4} = 1.37$ and P-value $= 0.3524$.

(c) No. The estimate of σ_β^2 is negative.

Chapter 15

2^k Factorial Experiments and Fractions

15.1 Either using Table 15.5 (e.g., $SSA = \frac{(-41+51-57-63+67+54-76+73)^2}{24} = 2.6667$) or running an analysis of variance, we can get the Sums of Squares for all the factorial effects.

$$SSA = 2.6667, \qquad SSB = 170.6667, \quad SSC = 104.1667, \qquad SS(AB) = 1.500.$$
$$SS(AC) = 42.6667, \quad SS(BC) = 0.0000, \quad SS(ABC) = 1.5000.$$

15.3 The AD and BC interaction plots are printed here. The AD plot varies with levels of C since the ACD interaction is significant, or with levels of B since ABD interaction is significant.

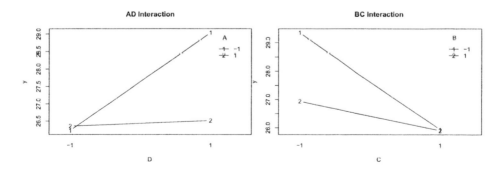

15.5 The ANOVA table is displayed.

Source of Variation	Degrees of Freedom	Computed f	P-value
A	1	9.98	0.0251
B	1	0.20	0.6707
C	1	6.54	0.0508
D	1	0.02	0.8863
AB	1	1.83	0.2338
AC	1	0.20	0.6707
AD	1	0.57	0.4859
BC	1	19.03	0.0073
BD	1	1.83	0.2338
CD	1	0.02	0.8863
Error	5		
Total	15		

One two-factor interaction BC, which is the interaction of Blade Speed and Condition of Nitrogen, is significant. As of the main effects, Mixing time (A) and Nitrogen Condition (C) are significant. Since BC is significant, the insignificant main effect B, the Blade Speed, cannot be declared insignificant. Interaction plots for BC at different levels of A are given here.

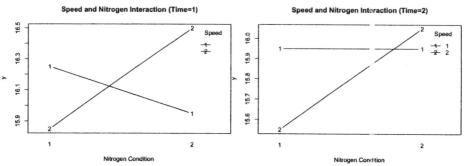

15.7 Both AD and BC interaction plots are shown in Exercise 15.3. Here is the interaction plot of AB.

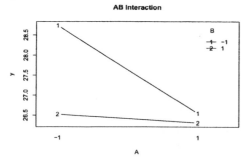

For AD, at the high level of A, Factor D essentially has no effect, but at the low level of A, D has a strong positive effect. For BC, at the low level of B, Factor C has a strong negative effect, but at the high level of B, the negative effect of C is not as pronounced. For AB, at the high level of B, A clearly has no effect. At the low level of B, A has a strong negative effect.

15.9 (a) The parameter estimates for x_1, x_2 and $x_1 x_2$ are given as follows.

Variable	Degrees of Freedom	Estimate	f	P-value
x_1	1	5.50	5.99	0.0039
x_2	1	−3.25	−3.54	0.0241
$x_1 x_2$	1	2.50	2.72	0.0529

(b) The coefficients of b_1, b_2, and b_{12} are $w_A/2$, $w_B/2$, and $w_{AB}/2$, respectively.

(c) The P-values are matched exactly.

15.11 (a) The effects are given here and it appears that B, C, and AC are all important.

A	B	C	AB	AC	BC	ABC
0.875	5.875	9.625	−3.375	−9.625	0.125	−1.125

(b) The ANOVA table is given.

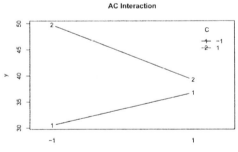

Source of Variation	Degrees of Freedom	Computed f	P-value
A	1	0.11	0.7528
B	1	4.79	0.0600
AB	1	12.86	0.0071
C	1	1.58	0.2440
AC	1	12.86	0.0071
BC	1	0.00	0.9640
ABC	1	0.18	0.6861
Error	8		
Total	15		

(c) Yes, they do agree.

(d) For the low level of Cutting Angle, C, Cutting Speed, A, has a positive effect on the life of a machine tool. When the Cutting Angle is large, Cutting Speed has a negative effect.

15.13 Here is the block arrangement.

	Block			Block			Block	
1		2	1		2	1		2
(1)		a	(1)		a	(1)		a
c		b	c		b	c		b
ab		ac	ab		ac	ab		ac
abc		bc	abc		bc	abc		bc

Replicate 1   Replicate 2   Replicate 3

AB Confounded AB Confounded AB Confounded

Analysis of Variance

Source of Variation	Degrees of Freedom
Blocks	5
A	1
B	1
C	1
AC	1
BC	1
ABC	1
Error	12
Total	23

15.15 $L_1 = \gamma_1 + \gamma_2 + \gamma_3$ and $L_2 = \gamma_1 + \gamma_2 + \gamma_4$. For treatment combination (1) we find $L_1 \,(\text{mod } 2) = 0$. For treatment combination a we find $L_1 \,(\text{mod } 2) = 1$ and $L_2 \,(\text{mod } 2) = 1$. After evaluating L_1 and L_2 for all 16 treatment combinations we obtain the following blocking scheme:

Block 1	Block 2	Block 3	Block 4
(1)	c	d	a
ab	abc	ac	b
acd	ad	bc	cd
bcd	bd	abd	abcd
$L_1 = 0$	$L_1 = 1$	$L_1 = 0$	$L_1 = 1$
$L_2 = 0$	$L_2 = 0$	$L_2 = 1$	$L_2 = 1$

Since $(ABC)(ABD) = A^2B^2CD = CD \,(\text{mod } 2)$, then CD is the other effect confounded.

15.17 $L_1 = \gamma_1 + \gamma_2 + \gamma_3$, $L_2 = \gamma_1 + \gamma_2$.

Block		Block		Block	
1	2	1	2	1	2
abc	*ab*	*abc*	*ab*	(1)	*a*
a	*ac*	*a*	*ac*	*c*	*b*
b	*bc*	*b*	*bc*	*ab*	*ac*
c	(1)	*c*	(1)	*abc*	*bc*
Rep 1		Rep 2		Rep 3	
ABC Confounded		*ABC* Confounded		*AB* Confounded	

For treatment combination (1) we find L_1 (mod 2) $= 0$ and L_2 (mod 2) $= 0$. For treatment combination a we find L_1 (mod 2) $= 1$ and L_2 (mod 2) $= 1$. Replicate 1 and Replicate 2 have $L_1 = 0$ in one block and $L_1 = 1$ in the other. Replicate 3 has $L_2 = 0$ in one block and $L_2 = 1$ in the other.

Analysis of Variance

Source of Variation	Degrees of Freedom
Blocks	5
A	1
B	1
C	1
AB	1'
AC	1
BC	1
ABC	1'
Error	11
Total	23

Relative information on $ABC = \frac{1}{3}$ and relative information on $AB = \frac{2}{3}$.

15.19 (a) One possible design would be:

Machine								
1	(1)	*ab*	*ce*	*abce*	*acd*	*bde*	*ade*	*bcd*
2	*a*	*b*	*ace*	*bce*	*cd*	*abde*	*de*	*abcd*
3	*c*	*abc*	*e*	*abe*	*ad*	*bcde*	*acde*	*bd*
4	*d*	*abd*	*cde*	*abcde*	*ac*	*be*	*ae*	*bc*

(b) ABD, CDE, and $ABCE$.

15.21 (a) The P-values of the regression coefficients are:

Parameter	Intercept	x_1	x_2	x_3	x_1x_2	x_1x_3	x_2x_3	$x_1x_2x_3$
P-value	< 0.0001	0.5054	0.0772	0.0570	0.0125	0.0205	0.7984	0.6161

and $s^2 = 0.57487$ with 4 degrees of freedom. So x_2, x_3, x_1x_2 and x_1x_3 are important in the model.

(b) $t = \dfrac{\bar{y}_f - \bar{y}_C}{\sqrt{s^2(1/n_f + 1/n_C)}} = \dfrac{52.075 - 49.275}{\sqrt{(0.57487)(1/8 + 1/4)}} = 6.0306$. Hence the P-value $= 0.0038$ for testing quadratic curvature. It is significant.

(c) Need one additional design point different from the original ones.

15.23 To estimate the quadratic terms, it might be good to add points in the middle of the edges. Hence $(-1,0)$, $(0,-1)$, $(1,0)$, and $(0,1)$ might be added.

15.25 (a) With BCD as the defining contrast, we have $L = \gamma_2 + \gamma_3 + \gamma_4$. The $\frac{1}{2}$ fraction corresponding to $L = 0$ (mod 2 is the principal block: $\{(1), a, bc, abc, bd, abd, cd, acd\}$.

(b) To obtain 2 blocks for the $\frac{1}{2}$ fraction the interaction ABC is confounded using $L = \gamma_1 + \gamma_2 + \gamma_3$:

Block 1	Block 2
(1)	a
bc	abc
abd	bd
acd	cd

(c) Using BCD as the defining contrast we have the following aliases:

$$A \equiv ABCD, \quad AB \equiv ACD, \quad B \equiv CD, \quad AC \equiv ABD,$$
$$C \equiv BD, \quad AD \equiv ABC, \quad D \equiv BC.$$

Since AD and ABC are confounded with blocks there are only 2 degrees of freedom for error from the unconfounded interactions.

Analysis of Variance

Source of Variation	Degrees of Freedom
Blocks	1
A	1
B	1
C	1
D	1
Error	2
Total	7

15.27 (a) With $ABCE$ and $ABDF$, and hence $(ABCE)(ABDF) = CDEF$ as the defining contrasts, we have

$$L_1 = \gamma_1 + \gamma_2 + \gamma_3 + \gamma_5, \qquad L_2 = \gamma_1 + \gamma_2 + \gamma_4 + \gamma_6.$$

The principal block, for which $L_1 = 0$, and $L_2 = 0$, is as follows:

$$\{(1), ab, acd, bcd, ce, abce, ade, bde, acf, bcf, df, abdf, aef, bef, cdef, abcdef\}.$$

(b) The aliases for each effect are obtained by multiplying each effect by the three defining contrasts and reducing the exponents modulo 2.

$$A \equiv BCE \equiv BDF \equiv ACDEF, \qquad B \equiv ACE \equiv ADF \equiv BCDEF,$$
$$C \equiv ABE \equiv ABCDF \equiv DEF, \qquad D \equiv ABCDE \equiv ABF \equiv CEF,$$
$$E \equiv ABC \equiv ABDEF \equiv CDF, \qquad F \equiv ABCEF \equiv ABD \equiv CDE,$$
$$AB \equiv CE \equiv DF \equiv ABCDEF, \qquad AC \equiv BE \equiv BCDF \equiv ADEF,$$
$$AD \equiv BCDE \equiv BF \equiv ACEF, \qquad AE \equiv BC \equiv BDEF \equiv ACDF,$$
$$AF \equiv BCEF \equiv BD \equiv ACDE, \qquad CD \equiv ABDE \equiv ABCF \equiv EF,$$
$$DE \equiv ABCD \equiv ABEF \equiv CF, \qquad BCD \equiv ADE \equiv ACF \equiv BEF,$$
$$DCF \equiv AEF \equiv ACD \equiv BDE,$$

Since E and F do not interact and all three-factor and higher interactions are negligible, we obtain the following ANOVA table:

Source of Variation	Degrees of Freedom
A	1
B	1
C	1
D	1
E	1
F	1
AB	1
AC	1
AD	1
BC	1
BD	1
CD	1
Error	3
Total	15

15.29 All two-factor interactions are aliased with each other. So, assuming that two-factor as well as higher order interactions are negligible, a test on the main effects is given in the ANOVA table.

Source of Variation	Degrees of Freedom	Mean Square	Computed f	P-value
A	1	6.125	5.81	0.0949
B	1	0.605	0.57	0.5036
C	1	4.805	4.56	0.1223
D	1	0.245	0.23	0.6626
Error	3	1.053		
Total	7			

Apparently no main effects is significant at level 0.05. Comparatively factors A and C are more significant than the other two. Note that the degrees of freedom on the error term is only 3, the test is not very powerful.

15.31 To get all main effects and two-way interactions in the model, this is a saturated design, with no degrees of freedom left for error. Hence, we first get all SS of these effects and pick the 2-way interactions with large SS values, which are AD, AE, BD and BE. An ANOVA table is obtained.

Source of Variation	Degrees of Freedom	Mean Square	Computed f	P-value
A	1	388,129.00	3,585.49	< 0.0001
B	1	277,202.25	2,560.76	< 0.0001
C	1	4,692.25	43.35	0.0006
D	1	9,702.25	89.63	< 0.0001
E	1	1,806.25	16.69	0.0065
AD	1	1,406.25	12.99	0.0113
AE	1	462.25	4.27	0.0843
BD	1	1,156.25	10.68	0.0171
BE	1	961.00	8.88	0.0247
Error	6	108.25		
Total	15			

All main effects, plus AD, BD and BE two-way interactions, are significant at 0.05 level.

15.33 Begin with a 2^3 with design points

$$\{(1), a, b, c, ab, ac, bc, abc\}.$$

Now, use the generator $D = AB$, $E = AC$, and $F = BC$. We have the following result:

$$\{def, af, be, cd, abd, ace, bcf, abcdef\}.$$

15.35 Here are all the aliases

$$A \equiv BD \equiv CE \equiv CDF \equiv BEF \equiv \ \equiv ABCF \equiv ADEF \equiv ABCDE;$$
$$B \equiv AD \equiv CF \equiv CDE \equiv AEF \equiv \ \equiv ABCE \equiv BDEF \equiv ABCDF;$$
$$C \equiv AE \equiv BF \equiv BDE \equiv ADF \equiv \ \equiv CDEF \equiv ABCD \equiv ABCEF;$$
$$D \equiv AB \equiv EF \equiv BCE \equiv ACF \equiv \ \equiv BCDF \equiv ACDE \equiv ABDEF;$$
$$E \equiv AC \equiv DF \equiv ABF \equiv BCD \equiv \ \equiv ABDE \equiv BCEF \equiv ACDEF;$$
$$F \equiv BC \equiv DE \equiv ACD \equiv ABE \equiv \ \equiv ACEF \equiv ABDF \equiv BCDEF.$$

15.37 When the variables are centered and scaled, the fitted model is

$$\hat{y} = 12.7519 + 4.7194x_1 + 0.8656x_2 - 1.4156x_3.$$

The lack-of-fit test results in an f-value of 81.58 with P-value < 0.0001. Hence, higher-order terms are needed in the model.

15.39 The defining contrasts are

$$AFG, \; CEFG, \; ACDF, \; BEG, \; BDFG, \; CDG, \; BCDE, \; ABCDEFG, \; DEF, \; ADEG.$$

Chapter 16

Nonparametric Statistics

16.1 The hypotheses

$$H_0 : \mu = 20 \text{ minutes}$$
$$H_1 : \tilde{\mu} > 20 \text{ minutes.}$$

$\alpha = 0.05$.

Test statistic: binomial variable X with $p = 1/2$.

Computations: Subtracting 20 from each observation and discarding the zeroes. We obtain the signs

$$- \quad + \quad + \quad - \quad + \quad + \quad - \quad + \quad + \quad +$$

for which $n = 10$ and $x = 7$. Therefore, the P-value is

$$P = P(X \geq 7 \mid p = 1/2) = \sum_{x=7}^{10} b(x; 10, 1/2)$$

$$= 1 - \sum_{x=0}^{6} b(x; 10, 1/2) = 1 - 0.8281 = 0.1719 > 0.05.$$

Decision: Do not reject H_0.

16.3 The hypotheses

$$H_0 : \tilde{\mu} = 2.5$$
$$H_1 : \tilde{\mu} \neq 2.5.$$

$\alpha = 0.05$.

Test statistic: binomial variable X with $p = 1/2$.

Computations: Replacing each value above and below 2.5 by the symbol "+" and "−", respectively. We obtain the sequence

$$- \quad - \quad - \quad - \quad - \quad - \quad - \quad + \quad + \quad - \quad + \quad - \quad - \quad - \quad - \quad -$$

for which $n = 16$, $x = 3$. Therefore, $\mu = np = (16)(0.5) = 8$ and $\sigma = \sqrt{(16)(0.5)(0.5)} = 2$. Hence $z = (3.5 - 8)/2 = -2.25$, and then

$$P = 2P(X \leq 3 \mid p = 1/2) \approx 2P(Z < -2.25) = (2)(0.0122) = 0.0244 < 0.05.$$

Decision: Reject H_0.

16.5 The hypotheses

$$H_0 : \tilde{\mu}_1 - \tilde{\mu}_2 = 4.5$$
$$H_1 : \tilde{\mu}_1 - \tilde{\mu}_2 < 4.5.$$

$\alpha = 0.05$.
Test statistic: binomial variable X with $p = 1/2$.
Computations: We have $n = 10$ and $x = 4$ plus signs. Therefore, the P-value is

$$P = P(X \leq 4 \mid p = 1/2) = \sum_{x=0}^{4} b(x; 10, 1/2) = 0.3770 > 0.05.$$

Decision: Do not reject H_0.

16.7 The hypotheses

$$H_0 : \tilde{\mu}_2 - \tilde{\mu}_1 = 8$$
$$H_1 : \tilde{\mu}_2 - \tilde{\mu}_1 < 8.$$

$\alpha = 0.05$.
Test statistic: binomial variable X with $p = 1/2$.
Computations: We have $n = 13$ and $x = 4$. Therefore, $\mu = np = (13)(1/2) = 6.5$ and $\sigma = \sqrt{(13)(1/2)(1/2)} = 1.803$. Hence, $z = (4.5 - 6.5)/1.803 = -1.11$, and then

$$P = P(X \geq 4 \mid p = 1/2) = P(Z < -1.11) = 0.1335 > 0.05.$$

Decision: Do not reject H_0.

16.9 The hypotheses

$$H_0 : \tilde{\mu} = 12$$
$$H_1 : \tilde{\mu} \neq 12.$$

$\alpha = 0.02$.
Critical region: $w_{\leq}20$ for $n = 15$.
Computations:

d_i	-3	1	-2	-1	6	4	1	2	-1	3	-3	1	2	-1	2
Rank	12	3.5	8.5	3.5	15	14	3.5	8.5	3.5	12	12	3.5	8.5	3.5	8.5

Now, $w_- = 43$ and $w_+ = 77$, so that $w = 43$.
Decision: Do not reject H_0.

16.11 The hypotheses

$$H_0 : \tilde{\mu}_1 - \tilde{\mu}_2 = 4.5$$
$$H_1 : \tilde{\mu}_1 - \tilde{\mu}_2 < 4.5.$$

$\alpha = 0.05$.
Critiral region: $w+ \leq 11$.
Computations:

Woman	1	2	3	4	5	6	7	8	9	10
d_i	−1.5	5.4	3.6	6.9	5.5	2.7	2.3	3.4	5.9	0.7
$d_i - d_0$	−6.0	0.9	−0.9	2.4	1.0	−1.8	−2.2	−1.1	1.4	−3.8
Rank	10	1.5	1.5	8	3	6	7	4	5	9

Therefore, $w_+ = 17.5$.
Decision: Do not reject H_0.

16.13 The hypotheses

$$H_0 : \tilde{\mu}_1 - \tilde{\mu}_2 = 8$$
$$H_1 : \tilde{\mu}_1 - \tilde{\mu}_2 < 8.$$

$\alpha = 0.05$.
Critiral region: $z < -1.645$.
Computations:

d_i	6	9	3	5	8	9	4	10
$d_i - d_0$	−2	1	−5	−3	0	1	−4	2
Rank	4.5	1.5	10.5	7.5	−	1.5	9	4.5
d_i	8	2	6	3	1	6	8	11
$d_i - d_0$	0	−6	−2	−5	−7	−2	0	3
Rank	−	12	4.5	10.5	13	4.5	−	7.5

Discarding zero differences, we have $w_1 = 15$, $n = 13$, $\mu_{W_+} - (13)(14)/4 = 45, 5$, and $\sigma_{W_+} = \sqrt{(13)(14)(27)/24} = 15.309$. Therefore, $z = (15 - 45.5)/14.309 = -2.13$
Decision: Reject H_0; the average increase is less than 8 points.

16.15 The hypotheses

$$H_0 : \tilde{\mu}_B = \tilde{\mu}_A$$
$$II_1 : \tilde{\mu}_B < \tilde{\mu}_A.$$

$\alpha = 0.05$.
Critiral region: $n_1 = 3$, $n_2 = 6$ so $u_1 \leq 2$.
Computations:

Original data	1	7	8	9	10	11	12	13	14
Rank	1	2*	3*	4	5*	6	7	8	9

Now $w_1 = 10$ and hence $u_1 = 10 - (3)(4)/2 = 4$

Decision: Do not reject H_0; the claim that the tar content of brand B cigarettes is lower than that of brand A is not statistically supported.

16.17 The hypotheses

$$H_0 : \tilde{\mu}_A = \tilde{\mu}_B$$
$$H_1 : \tilde{\mu}_A > \tilde{\mu}_B.$$

$\alpha = 0.01$.
Critiral region: $u_2 \leq 14$.
Computations:

Original data	3.8	4.0	4.2	4.3	4.5	4 5	4.6	4.8	4.9
Rank	1*	2*	3*	4*	5.5*	5.5*	7	8*	9*
Original Data	5.0	5.1	5.2	5.3	5.5	5 6	5.8	6.2	6.3
Rank	10	11	12	13	14	15	16	17	18

Now $w_2 = 50$ and hence $u_2 = 50 - (9)(10)/2 = 5$
Decision: Reject H_0; calculator A operates longer.

16.19 The hypotheses

$$H_0 : \tilde{\mu}_1 = \tilde{\mu}_2$$
$$H_1 : \tilde{\mu}_1 \neq \tilde{\mu}_2.$$

$\alpha = 0.05$.
Critiral region: $u \leq 5$.
Computations:

Original data	64	67	69	75	78	79	80	82	87	88	91	93
Rank	1	2	3*	4	5*	6	7*	8	9*	10	11*	12

Now $w_1 = 35$ and $w_2 = 43$. Therefore, $u_1 = 35 - (5)(6)/2 = 20$ and $u_2 = 43 - (7)(8)/2 = 15$, so that $u = 15$.
Decision: Do not reject H_0.

16.21 The hypotheses

$$H_0 : \text{Operating times for all three calculators are equal.}$$
$$H_1 : \text{Operating times are not all equal.}$$

$\alpha = 0.01$.
Critiral region: $h > \chi^2_{0.01} = 9.210$ with $v = 2$ degrees of freedom.
Computations:

Ranks for Calculators		
A	B	C
4	8.5	15
12	7	18
1	13	10
2	11	16
6	8.5	14
$r_1 = 25$	5	17
	3	$r_3 = 90$
	$r_2 = 56$	

Now $h = \frac{12}{(18)(19)}\left[\frac{25^2}{5} + \frac{56^2}{7} + \frac{90^2}{6}\right] - (3)(19) = 10.47$.

Decision: Reject H_0; the operating times for all three calculators are not equal.

16.23 The hypotheses

$$H_0 : \text{Sample is random.}$$

$$H_1 : \text{Sample is not random.}$$

$\alpha = 0.1$.

Test statistics: V, the total number of runs.

Computations: for the given sequence we obtain $n_1 = 5$, $n_2 = 10$, and $v = 7$. Therefore, from Table A.18, the P-value is

$$P = 2P(V \le 7 \text{ when } H_0 \text{ is true}) = (2)(0.455) = 0.910 > 0.1$$

Decision: Do not reject H_0; the sample is random.

16.25 The hypotheses

$$H_0 : \mu_A = \mu_B$$
$$H_1 : \mu_A > \mu_B.$$

$\alpha = 0.01$.

Test statistics: V, the total number of runs.

Computations: from Exercise 16.17 we can write the sequence

$$B \quad B \quad B \quad B \quad B \quad B \quad A \quad B \quad B \quad A \quad A \quad B \quad A \quad A \quad A \quad A \quad A \quad A$$

for which $n_1 = 9$, $n_2 = 9$, and $v = 6$. Therefore, the P-value is

$$P = P(V \le 6 \text{ when } H_0 \text{ is true}) = 0.044 > 0.01$$

Decision: Do not reject H_0.

16.27 The hypotheses

$$H_0 : \text{Sample is random.}$$
$$H_1 : \text{Sample is not random.}$$

$\alpha = 0.05$.

Critical region: $z < -1.96$ or $z > 1.96$.

Computations: we find $\bar{x} = 2.15$. Assigning "+" and "−" signs for observations above and below the median, respectively, we obtain $n_1 = 15$, $n_2 = 15$, and $v = 19$. Hence,

$$\mu_V = \frac{(2)(15)(15)}{30} + 1 = 16,$$
$$\sigma_V^2 = \frac{(2)(15)(15)[(2)(15)(15) - 15 - 15]}{(30^2)(29)} = 7.241,$$

which yields $\sigma_V = 2.691$. Therefore,

$$z = (19 - 16)/2.691 = 1.11.$$

Decision: Do not reject H_0.

16.29 $n = 24$, $1 - \alpha = 0.90$. From Table A.20, $1 - \gamma = 0.70$.

16.31 $n = 135$, $1 - \alpha = 0.95$. From Table A.21, $1 - \gamma = 0.995$.

16.33 (a) Using the following

Ranks			Ranks		
x	y	d	x	y	d
1	6	−5	14	12	2
2	1	1	15	2	13
3	16	−13	16	6	10
4	9.5	−5.5	17	13.5	3.5
5	18.5	−13.5	18	13.5	4.5
6	23	−17	19	16	3
7	8	−1	20	23	−3
8	3	5	21	23	−2
9	9.5	−0.5	22	23	−1
10	16	−6	23	18.5	4.5
11	4	7	24	23	1
12	20	−8	25	6	19
13	11	2			

we obtain $r_S = 1 - \frac{(6)(1586.5)}{(25)(625-1)} = 0.39$.

(b) The hypotheses

$$H_0 : \rho = 0$$
$$H_1 : \rho \neq 0$$

$\alpha = 0.05$.
Critical region: $r_S < -0.400$ or $r_s > 0.400$.
Decision: Do not reject H_0.

16.35 (a) We have the following table:

Weight	Chest Size	d_i	Weight	Chest Size	d_i	Weight	Chest Size	d_i
3	6	-3	1	1	0	8	8	0
9	9	0	4	2	2	7	3	4
2	4	-2	6	7	-1	5	5	0

$$r_S = 1 - \frac{(6)(34)}{(9)(80)} = 0.72.$$

(b) The hypotheses

$$H_0 : \rho = 0$$
$$H_1 : \rho > 0$$

$\alpha = 0.025$.
Critical region: $r_S > 0.683$.
Decision: Reject H_0 and claim $\rho > 0$.

16.37 (a) $\sum d^2 = 24$, $r_S = 1 - \frac{(6)(24)}{(8)(63)} = 0.71$.

(b) The hypotheses

$$H_0 : \rho = 0$$
$$H_1 : \rho > 0$$

$\alpha = 0.05$.
Critical region: $r_S > 0.643$.
Computations: $r_S = 0.71$.
Decision: Reject H_0, $\rho > 0$.

Chapter 17

Statistical Quality Control (Blank)

Due to the fact that the problems in this chapter are all review chapters, there are no solutions available for the problems in this chapter.

Chapter 18

Bayesian Statistics

18.1 For $p = 0.1, b(2; 2, 0.1) = \binom{2}{2}(0.1)^2 = 0.01$.
For $p = 0.2, b(2; 2, 0.2) = \binom{2}{2}(0.2)^2 = 0.04$. Denote by

$$A: \quad \text{number of defectives in our sample is 2;}$$
$$B_1: \quad \text{proportion of defective is } p = 0.1;$$
$$B_2: \quad \text{proportion of defective is } p = 0.2.$$

Then

$$P(B_1|A) = \frac{(0.6)(0.01)}{(0.6)(0.01) + (0.4)(0.04)} = 0.27,$$

and then by subtraction $P(B_2|A) = 1 - 0.27 = 0.73$. Therefore, the posterior distribution of p after observing A is

p	0.1	0.2
$\pi(p\|x = 2)$	0.27	0.73

for which we get $p^* = (0.1)(0.27) + (0.2)(0.73) = 0.173$.

18.3 (a) Let X = the number of drinks that overflow. Then

$$f(x|p) = b(x; 4, p) = \binom{4}{x}p^x(1 - p)^{4-x}, \quad \text{for } x = 0, 1, 2, 3, 4.$$

Since

$$f(1, p) = f(1|p)\pi(p) = 10\binom{4}{1}p(1 - p)^3 = 40p(1 - p)^3, \quad \text{for } 0.05 < p < 0.15,$$

then

$$g(1) = 40 \int_{0.05}^{0.15} p(1 - p)^3 \, dp = -2(1 - p)^4 \, (4p + 1)|_{0.05}^{0.15} = 0.2844,$$

and

$$\pi(p|x = 1) = 40p(1 - p)^3/0.2844.$$

(b) The Bayes estimator

$$p^* = \frac{40}{0.2844} \int_{0.05}^{0.15} p^2(1-p)^3 \, dp$$

$$= \frac{40}{(0.2844)(60)} p^3 \left(20 - 45p + 36p^2 - 10p^3\right)\Big|_{0.05}^{0.15} = 0.106.$$

18.5 $n = 10, \bar{x} = 9, \sigma = 0.8, \mu_0 = 8, \sigma_0 = 0.2$, and $z_{0.025} = 1.96$. So,

$$\mu_1 = \frac{(10)(9)(0.04) + (8)(0.64)}{(10)(0.04) + 0.64} = 8.3846, \quad \sigma_1 = \sqrt{\frac{(0.04)(0.64)}{(10)(0.04) + 0.64}} = 0.1569.$$

To calculate Bayes interval, we use $8.3846 \pm (1.96)(0.1569) = 8.3846 \pm 0.3075$ which yields $(8.0771, 8.6921)$. Hence, the probability that the population mean is between 8.0771 and 8.6921 is 95%.

18.7 (a) $P(71.8 < \mu < 73.4) = P\left(\frac{71.8-72}{\sqrt{5.76}} < Z < \frac{73.4-72}{\sqrt{5.76}}\right) = P(-0.08 < Z < 0.58) = 0.2509.$

(b) $n = 100, \bar{x} = 70, s^2 = 64, \mu_0 = 72$ and $\sigma_0^2 = 5.76$. Hence,

$$\mu_1 = \frac{(100)(70)(5.76) + (72)(64)}{(100)(5.76) + 64} = 70.2,$$

$$\sigma_1 = \sqrt{\frac{(5.76)(64)}{(100)(5.76) + 64}} = 0.759.$$

Hence, the 95% Bayes interval can be calculated as $70.2 \pm (1.96)(0.759)$ which yields $68.71 < \mu < 71.69$.

(c) $P(71.8 < \mu < 73.4) = P\left(\frac{71.8-70.2}{0.759} < Z < \frac{73.4-70.2}{0.759}\right) = P(2.11 < Z < 4.22) = 0.0174.$

18.9 Multiplying the likelihood function and the prior distribution together, we get the joint density function of θ as

$$f(t_1, t_2, \ldots, t_n, \theta) = 2\theta^n \exp\left[-\theta\left(\sum_{i=1}^{n} t_i + 2\right)\right], \quad \text{for } \theta > 0.$$

Then the marginal distribution of $(T_1, T_2, \ldots, T_n)$ is

$$g(t_1, t_2, \ldots, t_n) = 2 \int_0^\infty \theta^n \exp\left[-\theta\left(\sum_{i=1}^n t_i + 2\right)\right] d\theta$$

$$= \frac{2\Gamma(n+1)}{\left(\sum_{i=1}^n t_i + 2\right)^{n+1}} \int_0^\infty \frac{\theta^n \exp\left[-\theta\left(\sum_{i=1}^n t_i + 2\right)\right]}{\Gamma(n+1)\left(\sum_{i=1}^n t_i + 2\right)^{-(n+1)}} d\theta$$

$$- \frac{2\Gamma(n+1)}{\left(\sum_{i=1}^n t_i + 2\right)^{n+1}},$$

since the integrand in the last term constitutes a gamma density function with parameters $\alpha = n + 1$ and $\beta = 1/\left(\sum_{i=1}^n t_i + 2\right)$. Hence, the posterior distribution of θ is

$$\pi(\theta|t_1, \ldots, t_n) = \frac{f(t_1, \ldots, t_n, \theta)}{g(t_1, \ldots, t_n)} = \frac{\left(\sum_{i=1}^n t_i + 2\right)^{n+1}}{\Gamma(n+1)} \theta^n \exp\left[-\theta\left(\sum_{i=1}^n t_i + 2\right)\right],$$

for $\theta > 0$, which is a gamma distribution with parameters $\alpha = n + 1$ and $\beta = 1/\left(\sum_{i=1}^n t_i + 2\right)$.

18.11 The likelihood function of p is $\binom{x-1}{4} p^5 (1-p)^{x-5}$ and the prior distribution is $\pi(p) = 1$. Hence the posterior distribution of p is

$$\pi(p|x) = \frac{p^5(1-p)^{x-5}}{\int_0^1 p^5(1-p)^{x-5}\, dp} = \frac{\Gamma(x+2)}{\Gamma(6)\Gamma(x-4)} p^5 (1-p)^{x-5},$$

which is a Beta distribution with parameters $\alpha = 6$ and $\beta = x - 4$. Hence the Bayes estimator, under the squared-error loss, is $p^* = \frac{6}{x+2}$.